THE 50 GREATEST SCIENTISTS

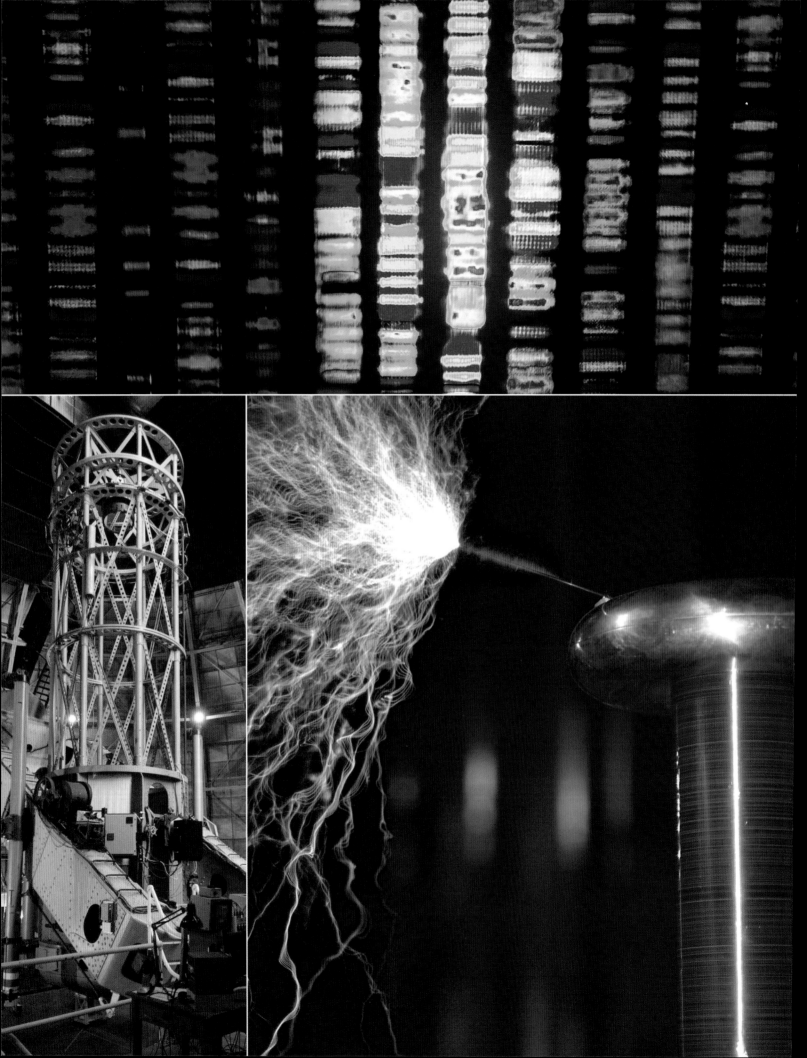

THE 50 GREATEST SCIENTISTS

THE PIONEERS WHO HAVE CHANGED OUR WORLD

JON BALCHIN

WITH CONTRIBUTIONS FROM
IAN FITZGERALD

SIRIUS

SIRIUS

This edition published in 2025 by Sirius Publishing, a division of
Arcturus Publishing Limited,
26/27 Bickels Yard, 151–153 Bermondsey Street,
London SE1 3HA

ISBN: 978-1-3988-3682-2
AD010857UK

Printed in China

Contents

Medical developments like Edward Jenner's discovery of vaccination have greatly benefited humanity.

INTRODUCTION

To be alive today is to be confronted by the products of science. Science has given us television, the internal combustion engine, the aeroplane, and the computer, to name but a few. Yet consumer products such as these are but one aspect of the benefits science can bring to mankind. Too often, for example, the field of medicine is overlooked in favour of more 'glamorous' fields, such as astrophysics or rocketry.

As recently as the nineteenth century, death from disease was an everyday occurrence. Both smallpox and polio killed millions until Edward Jenner made the simple yet life-changing discovery that milkmaids infected with cowpox were immune from smallpox, and Jonas Salk developed the polio vaccine. That both diseases continue to be killers in the modern world is due not to science, but to a tragic reluctance on the part of richer countries to share its benefits with their poorer counterparts.

Science has also produced less beneficial developments; the tank, machine gun, and atomic bomb, but science does achieve results, however morally questionable some of those results may be, and it is this which sets it apart from superstition, witchcraft and religion.

Important though the products of science may be, what is perhaps more significant is the scientific method itself, proceeding as it does from empirical observation to theory, to modification of theory in the light of further evidence.

We may still pray for rain, but we understand the physical causes of the weather, and to an extent can predict it; we no longer ascribe it to the actions of some unknowable deity and sacrifice our first-born in the hope of a favourable outcome.

This method contrasts with the previous means of discovering truth 'by authority', which claimed beliefs as true not on the basis of what was claimed, rather on the basis of who was making the claim.

Rejecting the notion of truth by authority, the scientists in this book observed the world around them, proposed theories to explain it, and modified these theories to account for further observations.

The road out from the darkness of superstition into the light of reason has not always been an easy one. When Vesalius dared to contradict the authority of Galen, he was abused as a liar and madman; the Montgolfier brothers' claims met only scepticism. Galileo and Copernicus both narrowly avoided following Giordano Bruno to the stake for proposing the heliocentric theory of the solar system, in opposition to accepted Church dogma. Yet they persevered, and in so doing, lit a beacon for the rest of humanity to follow.

The men and women who make up this book have blazed, in Bertrand Russell's poetic phrase, 'with all the noonday brightness of human genius'. How far the beacon they have lit will guide us, and how far science will yet progress, however, we shall leave to the next generation of scientists who will change the world.

The Montgolfier brothers' claims to have discovered a means of flight were met with unrelenting scepticism.

PYTHAGORAS

'NUMBER IS THE RULER OF FORMS AND IDEAS, AND THE CAUSE OF GODS AND DAEMONS.'

- Pythagoras

GREECE
c. 581–497 BCE

IDEAS AND INVENTIONS
Pythagoras' Theorem, the World as a Sphere, Harmonics

FIELDS
Mathematics, Physics, Astronomy

Opposite: Pythagoras.

The Pythagoreans may have made many of the discoveries usually attributed to Pythagoras himself.

Very little is known for certain about the life of this Greek mathematician and philosopher. We know that in 525 BCE he was taken prisoner by the Babylonians. Around 518 BCE he established his own academy at Croton, now Crotone, in southern Italy. This semi-religious, philosophical school gave birth to a number of disciples, known as the Pythagoreans – who may have actually made many of the mathematical discoveries usually credited to Pythagoras. Moreover, because of the reverence with which the originator of the brotherhood was treated by his followers and biographers, it is sometimes difficult to discern legend from fact.

It is fairly clear that Pythagoras himself did undertake practical experiments concerning the relationship between mathematics and music. It is thought he either attached different weights to a series of strings, or alternatively experimented with different string lengths, examining the mathematical relationship between the resultant notes when plucked and the weights or lengths applied. What he discovered was that simple, whole number relations, for example a string of one length and another of twice that length, produced harmonious tones. These observations ultimately led to the determination of musical scales as we know them today. Not only was this a momentous musical discovery, but was probably the first time a physical law had been mathematically expressed. As a result it began the science of mathematical physics.

This idea of a harmonious relationship between physical entities also enabled Pythagoras to conceptualize the world as a sphere, even if he had limited scientific basis at that time with which to back up his belief. For Pythagoras and his followers the idea of a 'perfect' mathematical interrelation between a globe moving in circles and the stars behaving similarly in a spherical universe (just as musical tones harmoniously danced around and depended on each other) seemed much more pleasing than Anaximander's cylindrical earth, or one composed of a flat disc. The view was so powerful that it inspired later Greek scholars, including Aristotle, to seek and ultimately find physical and mathematical evidence to reinforce the theory of the world as an orb.

When Pythagoras founded his school at Croton in Italy, one of its objectives was to further explore the relationship between the physical world and mathematics. Indeed, of the five key beliefs that the Pythagoreans held, one was dominant: the idea that 'all is number'. In other words, reality is at its fundamental level mathematical and that all physical things, like musical scales or the spherical earth and its companions the stars and the universe, are mathematically related. The experiments of the Pythagoreans led to numerous discoveries such as 'the

Pythagoras explored the relationship between mathematics and music.

Pythagoras was one of the first to conceive of the Earth as a sphere.

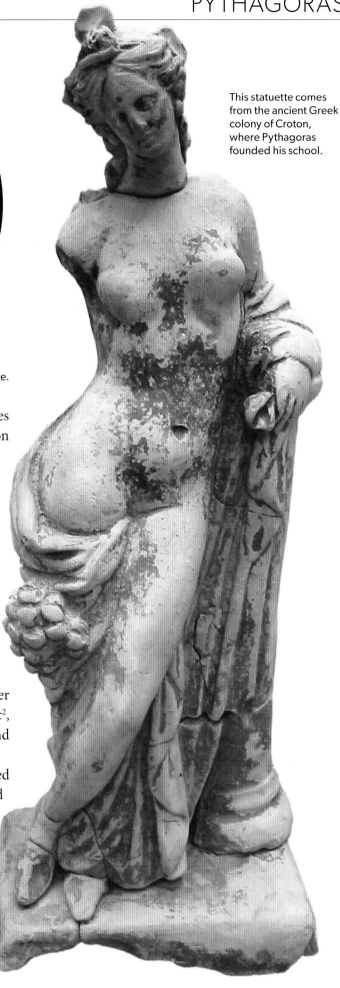

This statuette comes from the ancient Greek colony of Croton, where Pythagoras founded his school.

sum of a triangle's angles is equal to two right angles (180°). 'The sum of the interior angles in a polygon of n sides is equal to 2n-4 right angles' was another. Yet arguably their most important arithmetical discovery was that of irrational numbers. This came from the realization that the square root of two could not be expressed as a perfect fraction. This was a major blow to the Pythagorean idea of perfection and according to some accounts attempts were even made to try to conceal the discovery.

Pythagoras' famous Theorem was probably known to the Babylonians but Pythagoras may well have been the first to mathematically prove it. 'The square of the hypotenuse on a right-angled triangle is equal to the sum of the squares on the other two sides' can otherwise be expressed as $a^2+b^2=c^2$, where a and b are the shorter sides of the triangle and c is the hypotenuse.

It is perhaps ironic that Pythagoras is remembered today for his Theorem, the principles of which had previously been known for over a thousand years, and yet his more original discoveries are obscure. As the discoverer of the musical scale, in effect creating a rule book for the musical harmonies that we take for granted, it is arguable that this has had a much more profound impact on the history of the world than a simple, largely borrowed, mathematical formula.

HIPPOCRATES OF COS

GREECE
c. 460–377 BCE

IDEAS AND INVENTIONS
Hippocratic Oath, the Four Humours

FIELDS
Medicine

Below: Hippocrates.

'AS TO DISEASES, MAKE A HABIT OF TWO THINGS – TO HELP, OR AT LEAST TO DO NO HARM.'

- Hippocrates, *Epidemics*

Much of what is attributed to Hippocrates is contained within *The Hippocratic Collection*, a series of sixty to seventy medical texts written in the late fifth and early fourth century BCE. It is widely acknowledged, though, that Hippocrates himself could not have written many of these works, which is why precise details about his life and achievements remain unclear. Written over the period of a century and varying hugely in style and argument, it is thought they came from the medical school library of Cos, possibly put together in the first instance by the author to whom they later became attributed.

Given the sobriquet of the 'Great Physician' by Aristotle, Hippocrates is today more commonly referred to as the 'Father of Medicine'. Without question, Hippocrates of Cos, in spite of the limited factual details actually known about his life, helped lay the foundation stones of the science of medicine and greatly influenced its later development, even up to the present day.

For Hippocrates, disease and its treatment were entirely of this earth. He cast aside superstition and focused on the natural, in particular observing, recording and analysing the symptoms and passages of diseases. The prognosis of an illness was central to Hippocrates's approach to medicine, partly with a view to being able to avoid in future the circumstances which were perceived to have initiated the problems in the first place. The development of far-fetched cures or drugs, however, was not so important. What came from nature should, in Hippocrates's mind, be cured by nature; therefore rest, healthy diet, exercise, hygiene and air were prescribed for the treatment and prevention of illness.

Hippocrates set out the idea of the four humours which reigned as medical orthodoxy for nearly two millennia.

'Walking is a man's best medicine,' Hippocrates wrote.

He regarded the body as a single entity, or whole, and the key to remaining healthy lay in preserving the natural balance within this entity. The four 'humours' he believed influenced this equilibrium were blood, phlegm and yellow and black bile. When present in equal quantities, a healthy body would result. If one element became too dominant, however, then illness or disease would take over. The way to treat the problem would be by trying to undertake activities or eat foods which would stimulate the other humours, while at the same time attempting to restrain the dominant one, in order to restore the balance and, consequently, health.

Although this approach may still seem a little unscientific by today's standards of medicine, the fact that Hippocrates was prescribing such a natural, 'earthly' solution at all was a major advancement. Moreover, the concept and treatment of humours endured for the next two thousand years, certainly as far as the seventeenth century and in some aspects as far as the nineteenth. In addition, the answers he prescribed for healthy living such as diet and exercise are still 'good medicine' two thousand years later. Language introduced by Hippocrates also endures: an excess of black bile in Greek was known as 'melancholic', while someone with a too dominant phlegm humour was called 'phlegmatic'.

Ironically, Hippocrates may not have even written his own most enduring legacy. The Hippocratic Oath, probably penned by one of his followers, is a short passage constituting a code of conduct to which all physicians were henceforth obliged to pledge themselves. It outlines, amongst other things, the ethical responsibilities of the doctor to his patients and a commitment to patient confidentiality. It was an attempt to set physicians in the Hippocratic tradition apart from the spiritual and superstitious healers of their day. Such has been its durability that even students graduating from medical school today can still vow the Oath.

Before the time of Hippocrates there had been virtually no science at all in medicine. Disease was believed to be the punishment of the gods, Divine intervention came not from the natural, but the supernatural. The only 'treatment', therefore, also came from the supernatural: through magic, witchcraft, superstition or religious ritual.

Hippocrates confronted this notion head on, with a conviction remarkable given the age in which he lived. His approach brought the rational to the previously irrational and with it medicine strode into the age of reason. 'There are in fact two things,' said Hippocartes of Cos, 'science and opinion; the former begets knowledge, the latter ignorance.'

EUCLIDE Mathematicien
Renommé par ses Elements de Geometrie
il fleurifoit 319 avant l'ere Ch.

a Paris chez Daumont rue St Martin.

Cet ancien auteur, qui S'offre à tes regards ,
Est digne d'estre mis au rang des plus illustres
C'est lui, qui rassembla ces Elements Eparts .
Qu'on cite même Encor, depuis quatre cent lustres .

EUCLID

'THE LAWS OF NATURE ARE BUT THE MATHEMATICAL THOUGHTS OF GOD.'

- Euclid, *Elements*

GREECE
c. 330–260 BCE

IDEAS AND INVENTIONS
Euclidean Geometry

FIELDS
Mathematics

Left: Euclid.

I t is said that King Ptolemy I Soter of Egypt asked Euclid if it was possible to master geometry by a more direct route than reading his thirteen-volume definitive work on the topic. Euclid famously replied, 'There is no royal road to geometry, your Majesty.' Yet what Euclid had provided was one of the most majestic routes to the subject. It would go on to be revered for over two thousand years.

Euclid's legacy is well known, and yet, much of the life of the Greek mathematician remains a mystery. Although we possess extensive knowledge about the thoughts of many of the ancients, it is often the case that their lives and times are more obscure; this is certainly true in the case of Euclid. Although a name familiar to every schoolchild, almost nothing is known about his life, when and where he studied, or even when and where he was born and died: a true international man of mystery!

He probably studied under Plato at Athens and certainly spent most of his time in Alexandria where

he founded a mathematics academy. Whether all the works credited to him, including *Data*, *On Divisions*, the *Optics* and *Phaenomena*, were actually compiled solely by Euclid, or were produced with assistance from students at his school, remains unclear, but the impact of the texts is known to be great. In particular, *The Elements*, Euclid's masterwork on geometry, had a phenomenal influence on Western academic thinking. This is best illustrated by the suggestion that after the Bible, *The Elements* has probably been more studied, translated, and reprinted than any other book in history.

The reason for this is twofold: not just what Euclid said, but also the way that he said it. Indeed, the latter is arguably the more enduring of the two because it profoundly influenced the presentation of almost every future mathematical, scientific, theological and philosophical text, amongst others. The reason is because Euclid took a systematic approach to his writing, laying out a set of axioms (truths) at the beginning and constructing each proof of theorem which followed on the basis of the proven truths which had gone before. This logical, 'building block' method set the accepted academic precedent for proving knowledge and continues as a standard model today.

The compilation of knowledge that Euclid brought to the thirteen volumes of *The Elements* was so comprehensive and persuasive that it remained virtually unchanged and unchallenged as a teaching manual for over two millennia. Certainly, many of the theories outlined were not his; he was simply seeking to assimilate all geometric (and much other mathematical) knowledge into a single text. For example, the ideas of previous Greek mathematicians such as Eudoxus, Theaetetus and Pythagoras were all evident, though much of the systematic proof of theories, as well as other original contributions, was Euclid's. The first six of the thirteen volumes were concerned with plane geometry, for example laying out the basic principles of

Euclid holding his masterpiece, *The Elements*.

triangles, squares, rectangles and circles and any issues around these, as well as outlining other mathematical cornerstones, including Eudoxus's theory of proportion. The next four books looked at number theory, including the celebrated proof that there is an infinite number of prime numbers. The final three works focused on solid geometry.

Ironically, it is with some of the text's initial axioms that later mathematicians have found fault. The last axiom in particular has proved to be controversial. This 'parallel' axiom states that if a point lies outside a straight line, then only one straight line can be drawn through the point which never meets the other line in that plane (i.e. the parallel line). This was examined in the nineteenth century by the Romanian mathematician Janos Bolyai. Taking on his father's life work, he attempted to prove Euclid's parallel postulate, only to discover that, in fact, it was unprovable. This began a new school of thought and later, given further weight by Albert Einstein's belief that the geometry for space was also non-Euclidean, it was subsequently proved to be true.

Although the discoveries of the last two hundred years have shown time and space to be other than Euclidean under certain circumstances, this should not be seen to undermine his achievements. To have constructed *The Elements* in the manner he did, to have had an effect of such magnitude on the development of Western thought, and to have been accepted as the only authority on geometry for so long, (and for most practical purposes still attain such a status) is a profound legacy that few have equalled.

A Roman papyrus fragment from Egypt dating to c. 100 CE, showing a portion of Euclid's *Elements* from a school book. Euclid's *Elements* have formed the starting point for mathematical learning since they were first written.

ARCHIMEDES

'GIVE ME A PLACE TO STAND ON, AND I WILL MOVE THE EARTH.'

- Archimedes to the people of Syracuse.

GREECE
c. 287–212 BCE

IDEAS AND INVENTIONS
Archimedes' Principle, Integral Calculus, Archimedes' Screw

FIELDS
Mathematics, Engineering, Physics

Opposite: Archimedes.

While Archimedes may not have managed to move the earth itself, arranging for his patron King Heiron to move a ship by pushing a small lever was considered only a slightly less miraculous feat. With such audacious displays, along with his brilliance as an inventor, mechanical scientist and mathematician, it is no wonder Archimedes was so popular and highly regarded among his contemporaries.

It was not only his peers, however, who benefited from Archimedes' work. Many of his achievements are still with us today. First and foremost, Archimedes was an outstanding pure mathematician, 'usually considered to be one of the greatest mathematicians of all time,' according to the *Oxford Dictionary of Scientists*. He was, for example, the first to deduce that the volume of a sphere was $4\pi r^3 \times \frac{1}{3}$, where r is the radius. Other work in the same area, as outlined in his treatise *On the Sphere and Cylinder*, led him to deduce that a sphere's surface area can be worked out by multiplying that of its greatest circle by four; or, similarly, a sphere's volume is two-thirds that of its circumscribing cylinder. He calculated pi to be

ARCHIMEDES' PRINCIPLE

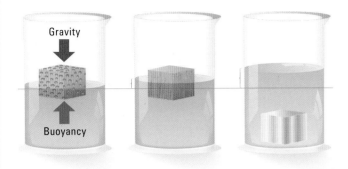

Archimedes' Principle – the buoyant force of an object in a liquid is equal to the weight of the fluid displaced by it.

approximately $^{22}/_7$, a figure that was widely used for the next 1,500 years.

Archimedes also discovered the principle that an object immersed in a liquid is buoyed or thrust upwards by a force equal to the weight of the fluid it displaces. The volume of the displaced liquid is the same as the volume of the immersed object. Legend has it that he discovered this when set a challenge by King Heiron to find out whether one of his crowns was made of pure gold or was a fake. While contemplating the problem Archimedes took a bath and noticed that the more he immersed his body in the water, the more the water overflowed from the tub. He realized that if he immersed the crown in a container of water and measured the water that overflowed he would know the volume of the crown. By obtaining a volume of pure gold equivalent to the volume of water displaced by the crown and then weighing both the crown and the gold, he could answer the King's question. On making this realization, Archimedes is said to have leapt from his tub and run naked along the street shouting 'Eureka!', 'I have found it!'

Indeed, it was the practical consequences of Archimedes' work which mattered more to his contemporaries and for which he became famous.

One such practical demonstration allowed King Heiron to move a ship with a single small lever – which in turn was connected to a series of other levers. Archimedes knew the experiment would work because he had already prepared a general theory of levers. Mathematically, he understood the relationship between the lever length, fulcrum position, the weight to be lifted and the force required

When Archimedes demonstrated his lever to King Heiron, he reportedly said 'Give me a place to stand on, and I will move the Earth.'

Archimedes supposedly created a giant mirror to reflect sunlight and burn Roman ships during the siege of Syracuse.

to move the weight. This meant he could successfully predict outcomes for any number of levers and objects to be lifted.

Likewise he came to understand and explain the principles behind the compound pulley, windlass, wedge and screw, as well as finding ways to determine the centre of gravity in objects.

Perhaps the most important inventions to his peers, however, were the devices created during the Roman siege of Syracuse in the second Punic War. He devised the Archimedes' Claw, a huge war machine designed to sink ships by grasping the prow and tipping them over, used in the defence of Syracuse. His machines helped repulse the Roman attack in 213 BCE, but a year later the Romans did eventually seize Syracuse, due to neglect of the defences, and Archimedes was killed by a Roman soldier while hard at work on mathematical diagrams. His last words are reputed to have been,

'Fellow, do not disturb my circles!' More than 100 years later, in 75 BCE, Archimedes' tomb was discovered and restored by the Roman statesman Cicero.

Archimedes left a remarkable legacy. As well as creating the science of hydrostatics (the study of the displacement of bodies in water), he also discovered the principles of static mechanics and pycnometry (the measurement of the volume or density of an object) and was a prolific inventor. Among other things, he invented compound pulley systems which enabled the lifting of enormous weights at a minimal expenditure of energy. He was also known as the 'father of integral calculus' for developing the method of exhaustion: an integral-like limiting process used to compute the area and volume of two-dimensional lamina and three-dimensional solids. Archimedes' reckonings were later used by, among others, Kepler, Fermat, Leibniz and Newton.

ZHANG HENG

CHINA
78–139 CE

IDEAS AND INVENTIONS
Seismograph, the figure of Pi, Armillary Sphere, Odometer

FIELDS
Mathematics, Astronomy, Geography

Opposite: Zhang Heng.

'THE EARTH IS LIKE THE YOLK OF THE EGG, LYING ALONE AT THE CENTRE. THE SKY IS LARGE AND THE EARTH SMALL.'

- Zhang Heng, *Hun yi*

The odometer, or south-pointing chariot, was one of Zhang Heng's many important inventions.

Western science is often credited with discoveries and inventions which have been observed in other cultures centuries before. This can be due to a lack of reliable records, difficulty in discerning fact from legend, problems in pinning down a finding to an individual or group, or frequently, simple ignorance. No such excuses exist for the work of Zhang Heng, whose life and achievements are well recorded, and whose major invention was created some 1,700 years before European scientists 'invented' the same thing.

Zhang, a Chinese scholar in the East Han Dynasty, was a man of many disciplines, including astronomy, mathematics and literature. He was born in Nanyang in central China in 78 CE. He studied in the major city of Chang'an (now Xi'an) and then the capital Luoyang.

Yet his greatest achievement was in geography, inspired by one of the duties assigned to him in the course of his work as Imperial Historian! China regularly suffered from earthquakes and as part of his job Zhang was required to record when and where they occurred. Rather than accept the common superstition that the quakes were punishment from angry gods, Zhang believed that if he took a scientific approach to noting data about tremors, the Dynasty would be better equipped to predict, prepare for and deal with them. To this end he devised the world's first seismograph, an invention he named Di Dong Yi, or 'Earth Motion Instrument'.

The seismograph was large, at almost two metres in diameter, and made out of bronze. Eight thin copper rods were attached to a central shaft at one end and to a corresponding number of dragons' heads at the other. These heads pointed in the eight major directions of a compass (north, north-east, east, south-east and so on), and each contained a copper ball in its mouth. Underneath each dragon was an open-mouthed copper frog. When a tremor occurred, the copper ball fell out of the mouth of the dragon nearest to the direction from which the earthquake came and into the frog's mouth, which in turn rang a bell alerting the royal household. A story is recorded that in 138 CE a copper ball had dropped to the west. Zhang recounted his finding to the emperor, but for two days nothing unusual happened and there were no reports of activity elsewhere. Sceptics were left to question the validity of Zhang's machine. Finally, though, messengers arrived on horseback reporting a severe earthquake 500 kilometres (310 miles) to the west. Zhang was vindicated.

Fabled to be a man with intense powers of concentration, Zhang was also able to employ his

A replica of Zhang Heng's seismoscope. It accurately predicted an earthquake in 138 CE.

An armillary sphere. In 117 CE, Zhang Heng made the world's first water-powered armillary sphere to help his astronomical calculations.

abilities to excellent effect in astronomy. Through his observations he correctly deduced that the sun caused the illumination of the moon, and that lunar eclipses were caused by the earth's shadow passing over its surface. He mapped the night sky in fine detail, recording 2,500 'brightly shining' stars in 124 constellations, 320 of which were named. He estimated that in total, including the 'very small', there were 11,520 stars. In addition, Zhang wrote a number of books on astronomy, the most famous being *Ling Xian*. In another, *Hun yi*, he outlined his perception of the universe and the earth's position within it. 'The sky is like a hen's egg,' he wrote, 'and is as round as a crossbow pellet; the earth is like the yolk of the egg, lying alone at the centre. The sky is large and the earth is small.'

Zhang Heng then, in common with his Greek predecessors, believed that the earth was spherical and at the centre of the universe. This drove him to create possibly the first three-dimensional model of the cosmos: a bronze celestial orb which turned by water-power. Each year, making a single complete rotation, it showed how the stars' positions changed.

Zhang undertook other work which had a lasting impact. He improved the previous figure of π from 3, the traditional figure in use by the Chinese, to √10 or 3.162, closer to the number of 3.142 used today. Zhang also performed calculations involving time, notably correcting the Chinese calendar in 123 CE to harmonize it with the seasons. Zhang's seismograph is recognized by the world as an instrument that was well ahead of its time. To this day no one has been able to reproduce it. He constructed the first accurate odometer, or 'mileage cart'. He was also considered one of the four great painters of the era. Zhang also produced over twenty famous literary works – he was a true polymath and left a remarkable legacy.

GALEN
OF PERGAMUM

'THE MIND'S INCLINATION FOLLOWS THE BODY'S TEMPERATURE.'

- attributed to Galen

ROMAN EMPIRE
130–201 CE

IDEAS AND INVENTIONS
Four Humours, Cardiovascular System

FIELDS
Anatomy, Medicine

Below: Galen.

Galen's fame comes not from any single achievement, but for the sheer volume of medical thought which he presented. His works on medical science became accepted as the only authority on the subject for the following 1,400 years.

The question, therefore, is why? Some commentaries suggest the answer is simply because Galen's studies were so all-encompassing that there was very little left for those following him to dispute. Another is the readiness with which the Arab, Christian and Jewish authorities accepted his work, lending it a weight which might have made it difficult for others to challenge. A third explanation could be that Galen not only incorporated the results of his own findings in his texts, but also compiled the best of all other medical knowledge that had gone before him into a single collection, such as that of Hippocrates, for example. In particular, Galen readily adopted Hippocrates' 'four humours' approach to the body, and this was one of the main reasons it endured for so long.

That is not to say Galen was at all lacking in original material and thinking. He was meticulous and methodical in his approach to his own medical investigations, above all in anatomy. Many important dignitaries came to the shrine of Asklepios, the god of healing in Galen's home town, to seek cures for ailments. Thus Galen was able to observe first hand

Galen and his fellow physicians, from the *Codex of Vienna Dioscurides*, c. 512 CE.

the symptoms and treatments of diseases. After spells in Smyrna (now Izmir), Corinth, and Alexandria studying both philosophy and medicine, which he considered inextricably linked, and including work on the dissection of animals, he returned to Pergamum in 157. There he took up a four-year appointment as a physician to gladiators, giving him further first hand experience in practical anatomical medicine.

All of this was excellent preparation for his transfer to Rome in 161. Here he spent most of the rest of his career and became the esteemed physician to emperors Marcus Aurelius, Lucius Verus, Commodus and Septimius Severus. This position not only brought him prestige, but it allowed him the freedom to undertake detailed research and dissection in the quest for the improved knowledge it provided.

Galen was not permitted to scrutinize human cadavers, so he dissected animals, predominantly Barbary apes because of the characteristics they shared with man. His most influential conclusions concerned the central operation of the human body. Sadly they were only influential in that they limited the search for accurate information for the next millennium and a half.

Temples to Asklepios, the god of healing, like this one on the Greek island of Kos, also served as early medical institutions.

Galen believed that blood was formulated in the liver, the source of natural spirit. In turn, this organ was nourished by the contents of the stomach which was transported to it. Veins from the liver carried blood to the extremes of the body where it was turned into flesh and 'used up', thus requiring more food on a daily basis to be converted into blood. Some of this blood passed through the heart's right ventricle, then seeped through to the left ventricle and mixed with air from the lungs, providing vital spirit which regulated the body's heat and blood flow. Using the arteries, a portion of this blood was then transported to the brain where it blended with animal spirit. This created movement and the senses. The combination of these three spirits managed the body and contributed to the make up of the soul. It

was for this reason that Galen missed the idea of a single, integrated system of the circulation of the blood, a result which was not conclusively proved until 1628 by William Harvey.

Although some of Galen's deductions were wrong, his surviving 129 volumes are a phenomenal contribution to his subject and offered a platform from which Renaissance physicians could begin their critical progress. It was Galen who first introduced the notion of experimentation to medicine. Many of the anatomical errors made by Galen were due to the fact that he could only operate on animals – human dissections were out of favour at the time. Galen became a doctor supposedly because his father had a dream in which Asklepios, the god of healing, appeared to him.

Galen served as the physician to the Roman Emperor Marcus Aurelius.

AVICENNA

'NO KNOWLEDGE IS ACQUIRED SAVE THROUGH THE STUDY OF ITS CAUSES AND BEGINNINGS.'

- Avicenna

PERSIA
980–1037

IDEAS AND INVENTIONS
*The Canon of Medicine,
The Book of Healing,*
psychotherapy, the speed of
light

FIELDS
Medicine, Astronomy,
Geometry, Philosophy

Opposite: Avicenna.

Born in present-day Uzbekistan in 980, in what was then the Sunni Muslim Samanid Empire stretching from Persia across Central Asia, Avicenna was the son of a high-ranking government official. His given name was Ibn Sina, Latinized as Avicenna, and he was raised in the imperial capital of Bukhara. An intellectually precocious child, he was educated in arithmetic, law, medicine, philosophy, and astronomy, and was said to have memorized the Quran by the age of 10. At 17 he was already one of the physicians attending to the Samanid ruler, Nuh II; later, like his father, he became an imperial administrator.

When the Samanid Empire fell to Turkish invaders known as the Qarakhanids in 999, Avicenna moved to the city of Gurganj, in modern Turkmenistan, a great seat of learning that attracted scholars from across the Qarakhanid Empire. By 1014 he had relocated to Ray, a city controlled by the Buyid Dynasty whose territory encompassed areas of modern Iran and Iraq. There, Avicenna became court physician to the Buyid ruler, or *amir*, Majd al-Dawla, and his powerful mother, Sayyid Shirin.

More political postings followed until, in 1021, he became a kind of court philosopher-scientist to Ala al-Dawla Muhammad, ruler of the great Silk Road city of Isfahan. Avicenna remained one of Ala al-Dawla's most trusted advisors until his death from colic in 1037 while on military campaign with his ruler.

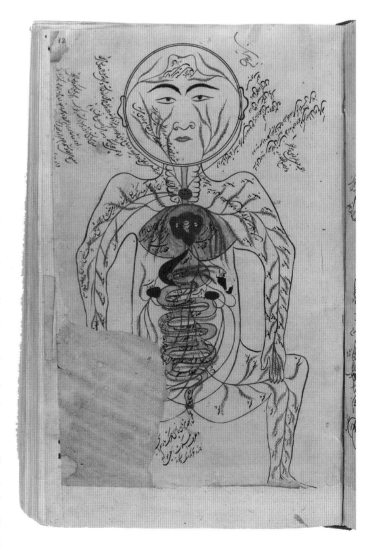

An illustration from Avicenna's *Canon of Medicine*.

THE ISLAMIC GOLDEN AGE

Avicenna lived in the middle of the Islamic Golden Age, an era from the 9th to the 14th centuries when Muslim-ruled cities such as Baghdad, Cairo, Cordoba, Gurganj and Isfahan became intellectual hotspots. While academic life flourished in the Islamic world, the Christian West was in turmoil for much of this period. The collapse of the Roman Empire at the end of the 5th century created a huge power vacuum across Europe that would take hundreds of years and countless destructive wars to resolve. Without the endeavours of Islamic scholars such as Avicenna, Averroës, Albucasis and Avenzoar, many of the great works of Classical figures such as Plato, Aristotle, Hippocrates and Archimedes would have disappeared into the culture-free abyss that was Europe's Dark Ages. By preserving, studying, and elaborating on the science, medicine, philosophy, mathematics, drama, art and architecture of the ancients, Avicenna and other Islamic intellectuals ensured the survival of priceless knowledge that would otherwise have been lost.

An illustration of the House of Wisdom in Baghdad. Founded in the late 8th century, it was the centre of Islamic intellectual life until its destruction by the Mongols in 1258.

Of his extensive works on subjects as diverse as geometry, astronomy, physics, philosophy, philology and music, Avicenna is best remembered as a physician who made significant advances in medical knowledge. His great writings were the *Book of Healing* and the *Canon of Medicine*, two multi-volume encyclopedias that became the basis for physicians in both the Islamic and Christian worlds until the 18th century. Avicenna's achievements as both a theoretical scientist and a working doctor are deeply impressive. He was the first person to identify that diseases such as measles, anthrax (which he called 'Persian Fire'), and tuberculosis were contagious, recognized the importance of the pulse as a diagnostic tool, and was an early advocate of using alcohol as an antiseptic. He advocated in his writings the use of forceps-like instruments to assist in childbirth – though it would be another 600 years before the physician Peter Chamberlain would invent them in the late 1600s.

An early herbalist, he recommended treating diseases and illness by natural means where possible and emphasized the importance of a healthy diet. As a profound thinker and student of psychology, he was one of the first people to link disease with mental health.

Avicenna was a great categorizer. Of his more than 400 written works, of which 240 survive, many of his medical studies are replete with lists and tables that still influence how physicians think about and present their work. Linked to this, Avicenna was an early champion of observation- and evidence-based medicine, along with drug-trialling. This allowed him to usefully group together types of illnesses and treatments scientifically. It is difficult to overstate the influence of Avicenna's empirical but structured and systematic approach to medicine.

Behind the well-deserved Great Man of Science persona Avicenna was, according to his

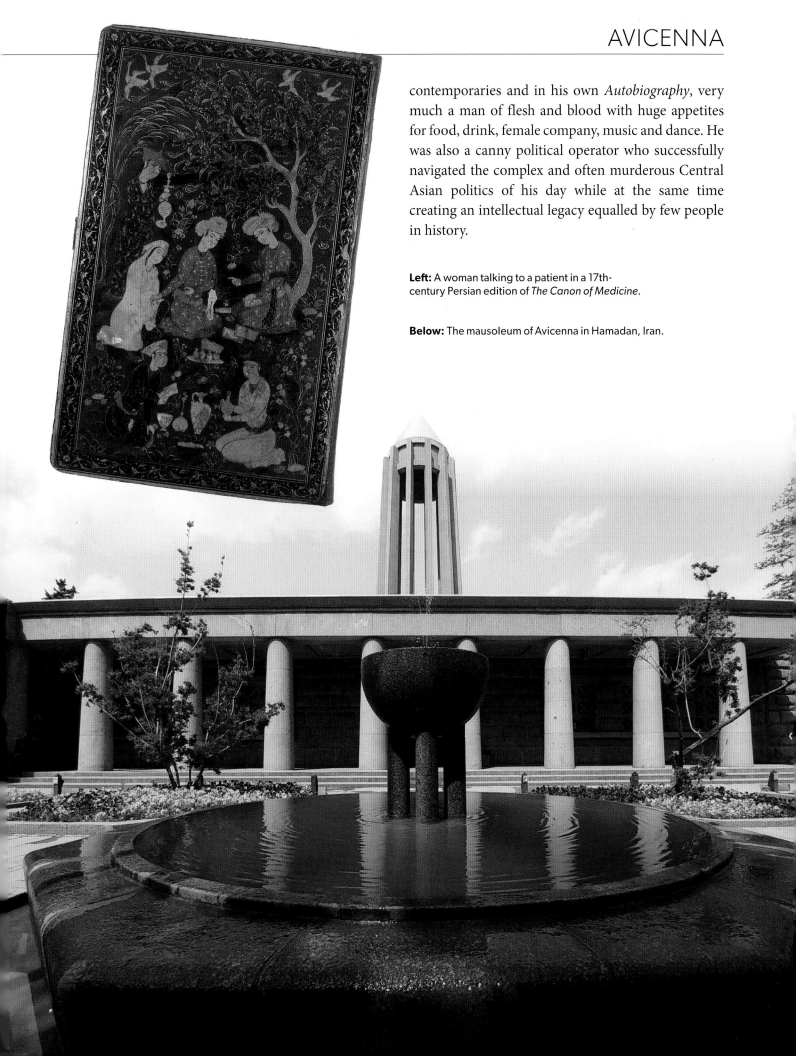

contemporaries and in his own *Autobiography*, very much a man of flesh and blood with huge appetites for food, drink, female company, music and dance. He was also a canny political operator who successfully navigated the complex and often murderous Central Asian politics of his day while at the same time creating an intellectual legacy equalled by few people in history.

Left: A woman talking to a patient in a 17th-century Persian edition of *The Canon of Medicine*.

Below: The mausoleum of Avicenna in Hamadan, Iran.

NON PAREM PAVLO GRATIÃ · REQVIRO
VENIAM PETRI NEQ. POSCO, SED QVAM
IN CRVCIS LIGNO DEDERAS LATRONI
SEDVLVS ORO

NICOLAUS COPERNICUS

POLAND
1473–1543

IDEAS AND INVENTIONS
Heliocentric model of the universe

FIELDS
Astronomy

Opposite: Nicolaus Copernicus.

'FINALLY WE SHALL PLACE THE SUN HIMSELF AT THE CENTRE OF THE UNIVERSE.'

- Copernicus

For all the impact the idea the planets might revolve around the sun, not the earth, would have on astronomy and science, arguably its biggest challenge would be to religion. The explanation of an earth inhabited by human beings, made in God's image as the most superior of all creatures, at the centre of a cosmos around which everything else revolved, suited the Christian Church's interpretation of the universe and mankind's position within it. It was a concept which dated back to Aristotle, was given observational legitimacy by Ptolemy and authority by Christendom. The Catholic religion still opposed the heliocentric model of planetary motion nearly three centuries after it was first published. And yet ironically its author, Nicolaus Copernicus, was himself a man of the Church.

Indeed, it was Copernicus' faith which had led him to question Ptolemy's accepted geocentric model of the universe in the first place. Why would God create a hugely complicated system of equants, epicycles and eccentrics, as Ptolemy had proposed, to explain the planets' motion around the earth when it would be far more simple, logical and graceful to have them all revolving around the sun? It was a theory Copernicus spent many years contemplating while studying in Krakow and then Italy, and continued to develop as he returned to Poland to take up a post as canon in Frauenberg Cathedral. He even used his position within the Church to quite literally advance his studies, using a cathedral tower to quietly and solitarily observe the stars.

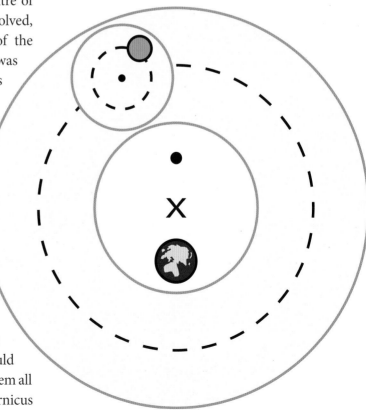

Ptolemy's model of the universe required a complex series of epicycles to explain the motion of the planets.

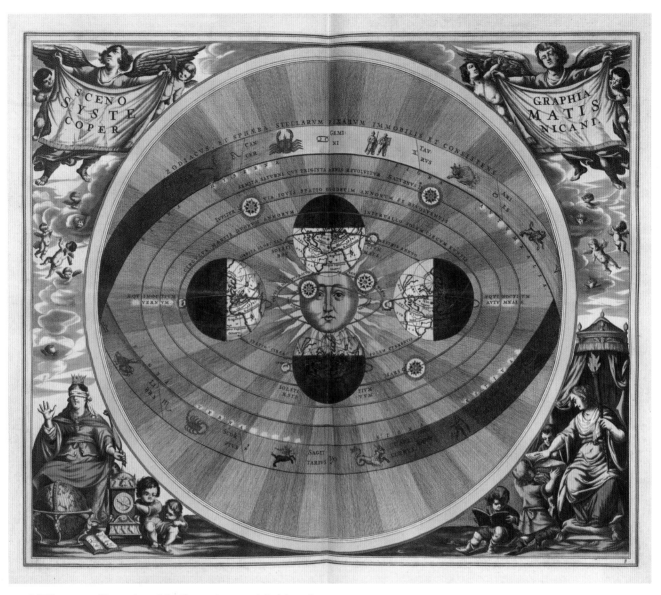

A 17th-century illustration of the Copernican model of the solar system.

Gradually Copernicus became more convinced of his proposition that a fixed sun was at the centre of planetary motion, with the earth rotating around it once a year. Between 1510 and 1514 he drafted *Commentariolus*, his initial exposition of the theory. In order to have any credence, the idea also required that the earth itself was not fixed in position as had previously been thought, but revolved on its axis once every twenty-four hours. This would also explain the apparent movement of the stars and sun across the sky. Perhaps because of his position within the Church, fearing a backlash, or perhaps because he was a perfectionist and recognized that his ideas were not fully developed, Copernicus resisted publishing *Commentariolus*, circulating it instead only among friends.

Copernicus continued to work on his ideas for the next twenty years and though his final work was largely completed by 1530 he continued to resist pleas by his friends to publish. Word of Copernicus' theories was already spreading across Europe and it is thought that even the Pope himself knew of them but offered no initial resistance to the idea of a heliocentric model. Indeed, it was not until 1616 that the Church banned the text Copernicus eventually published for its 'blasphemous' content, although that sanction subsequently remained in place until 1835, long after the 'Copernican system' had been widely accepted by most others.

On The Revolutions of Celestial Spheres was finally published in 1543. But as powerful and revolutionary

Jan Matejko's painting Astronomer Copernicus, or Conversation with God. Copernicus struggled with presenting his ideas to a wider audience for fear of backlash from the Church.

A page from the manuscript of *De Revolutionibus*.

as Copernicus' theories were, the text was rejected by many academics. This was partially because the author had undermined the simplicity of his initial ideas by clinging onto the Aristotelian belief that planetary motion took place in perfect circles. As we now know this not to be true, it meant Copernicus had been forced to introduce his own system of epicycles and other complex motions to fit in with observational evidence, thereby producing as equally complicated an explanation as the geocentric one he had initially rejected for its lack of simplicity. It was not until Johannes Kepler offered the solution that the planets moved in an elliptical, not circular, motion in 1609 that the simplicity Copernicus had been seeking was offered and the rest of his model could be vindicated.

Copernicus was brought up by his maternal uncle Lucas, the Bishop of Ermeland, and took a doctorate in canon law at the University of Ferrara in 1503. By this time he had become a canon of Frauenburg. Throughout his life Copernicus struggled to come to terms with the conflict between his mathematics and his religious faith. Indeed, one of the main reasons he did not publish his works was through fear of contradicting the Bible.

ANDREAS VESALIUS

BELGIUM
1514–1564

IDEAS AND INVENTIONS
Dissection, Vascular and Circulatory Systems, Nervous System

FIELDS
Anatomy, Medicine

Opposite: Andreas Vesalius.

'I COULD HAVE DONE NOTHING MORE WORTHWHILE THAN TO GIVE A NEW DESCRIPTION OF THE WHOLE HUMAN BODY.'

- Andreas Vesalius

It takes a brave person to challenge the accepted authority on any subject, especially one which has endured without dispute for some 1,400 years, and more especially when the person raising the objection is only twenty-eight years old and has only relatively recently graduated. That is just the task Andreas Vesalius took upon himself, however. For many of his contemporaries, there was nothing about this confrontation to consider as 'brave': instead, they described him as anything from a liar to a madman.

The authority Vesalius dared to challenge was that of Galen, the celebrated Roman physician who wrote what had been considered the definitive work on human anatomy. Such was his clout that when dissections of humans began to be permitted in Europe from the fourteenth century for research and tuition purposes, lecturers would simply read directly from Galen as the cadaver was cut by a butcher or assistant. Yet what was somehow lost sight of in all of this reverence was the fact that Galen himself had never actually dissected a human body, forbidden as this had been by Roman religious laws. Academics before Vesalius, however, still considered Galen as the authority on the subject, with any advance on his texts regarded as impossible.

Vesalius' approach was completely different. Born and raised in Belgium, to a family with a distinguished background as doctors to royalty, Vesalius was a keen dissector of animals from a young age. He went on to study medicine at institutions around Europe, notably the universities of Louvain, Paris and then Padua, where he was appointed Professor of Anatomy and Surgery at the age of 24. He insisted on performing the dissection of human bodies himself during lectures to students, rejecting the traditional clean-handed, textbook method.

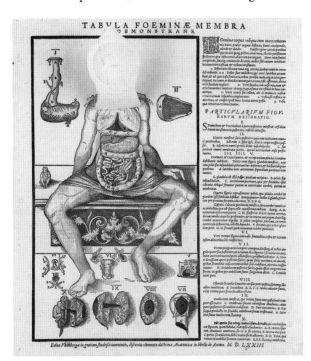

An anatomical fugitive sheet from 1573. These sheets, based on the pioneering work of Vesalius, allowed readers to lift flaps to reveal the underlying anatomy of a figure.

Although schooled in the Galenic tradition like all other medical students, Vesalius began questioning its teachings towards the end of the 1530s. From 1540 onwards, having been granted an ample number of human corpses to dissect, mostly from the local executioners, Vesalius was convinced. Galen's findings, he argued, did not reflect the human anatomy, but that of apes. This had led to numerous errors based on assumptions Galen had made on similarities between the two.

In 1543, Vesalius published his masterwork *De Humani Corporis Fabrica Libri Septem* or *The Seven Books on the Structure of the Human Body*. It was the first definitive work on human anatomy actually based on the results of methodical dissections of humans and, as such, was the most accurate work on the subject ever written. Furthermore, it was beautifully and clearly illustrated with woodcut drawings, probably prepared at the artist Titian's studios, and was excellently structured and organized. Its publication outdated all that had gone before and the text became the guide upon which future teachers would base their lectures. It was some time before its wisdom was widely accepted, however, due to the hostility which Vesalius often endured for challenging Galen. For example, Vesalius stated he could find no evidence for the 'pores' in the heart which allowed blood to seep from the right to the left ventricle, a key foundation of the Galenic tradition and one which was resolutely defended by many of his contemporaries.

Vesalius spent much of the rest of his life after the *Fabrica* in the service of kings, firstly as the physician to Charles V, the Holy Roman Emperor, then to Phillip II of Spain. He left Spain in 1564 on a pilgrimage to Jerusalem, but died on the return journey.

In spite of his premature death, Vesalius left behind a revolutionary legacy to anatomy students. It was only after his publications that both anatomy and medicine in general were first treated as sciences in their own right. By his reasoned critical approach to Galen, he had broken the reverence ascribed to the former 'master' and created a model for independent, rational investigation for his successors in the development of medical science.

Vesalius also changed the organization of the medical school classroom, and actively encouraged the participation of medical students in dissection lectures.

Left: A dissection drawing from *De Humani Corporis Fabrica* showing an accurate and detailed map of the muscles in the human body.

Opposite: Vesalius served as physician to the Holy Roman Emperor Charles V.

GALILEO
GALILEI

'I DISCOVERED IN THE HEAVENS MANY THINGS THAT HAD NOT BEEN SEEN BEFORE OUR OWN AGE.'

- Galileo Galilei

ITALY
1564–1642

IDEAS AND INVENTIONS
Momentum, Harmonic Motion, Hydrostatic Balance, Galilean Telescope, Heliocentric Model of the Solar System

FIELDS
Mathematics, Astronomy, Physics

Opposite: Galileo Galilei.

In both his life and through the imprisonment which he was forced to endure in the years leading up to his death, Galileo more than any other figure personified the optimism and struggle of the scientific revolution. He was responsible for a series of discoveries which would change our understanding of the world, while struggling against a society dominated by religious dogma, bent on suppressing his radical ideas.

Although he was initially encouraged to study medicine, Galileo's passion was mathematics, and it was his belief in this subject which underpinned all of his work. One of his most significant contributions was not least his application of mathematics to the science of mechanics, forging the modern approach to experimental and mathematical physics. He would take a problem, break it down into a series of simple parts, experiment on those parts, and then analyse the results until he could describe them in a series of mathematical expressions.

One of the areas in which Galileo had most success with this method was in explaining the rules of motion. In particular, the Italian rejected many of the Aristotelian explanations of physics which had largely endured to his day. One example was Aristotle's view that heavy objects fall towards earth faster than light ones. Through repeated experiments rolling different weighted balls down a slope (and, legend has it, dropping them from the top of the leaning tower of Pisa!), he found that they actually fell at the same rate. This led to his uniform theory of acceleration for falling bodies, which contended that in a vacuum all objects would accelerate at exactly the same rate towards earth, later proved to be true. Galileo also contradicted Aristotle in another area of motion by contending that a thrown stone had two forces acting upon it at the same time; one which we now know as 'momentum' pushing it horizontally, and another pushing downwards upon it, which we now know as 'gravity'. Galileo's work in these areas would prove vital to Isaac Newton's later discoveries.

Galileo rolled weighted balls down a slope to discover that they fell at the same speed, regardless of mass.

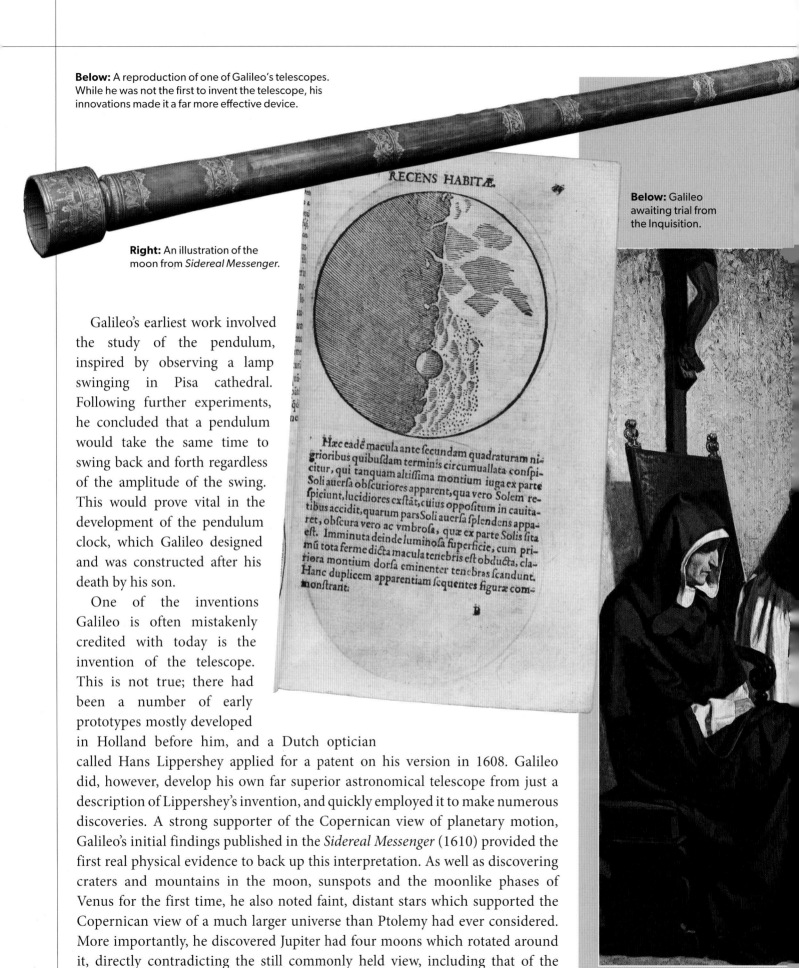

Below: A reproduction of one of Galileo's telescopes. While he was not the first to invent the telescope, his innovations made it a far more effective device.

Right: An illustration of the moon from *Sidereal Messenger*.

Below: Galileo awaiting trial from the Inquisition.

RECENS HABITÆ.

Hæc eadē macula ante secundam quadraturam nigrioribus quibusdam terminis circumuallata conspicitur, qui tanquam altissima montium iuga ex parte Soli auersa obscuriores apparent, qua vero Solem respiciunt, lucidiores exstat, cuius oppositum in cauitatibus accidit, quarum pars Soli auersa splendens apparet, obscura vero ac vmbrosa, quæ ex parte Solis sita est. Imminuta deinde luminosa superficie, cum primū tota ferme dicta macula tenebris est obducta, clariora montium dorsa eminenter tenebras scandunt. Hanc duplicem apparentiam sequentes figuræ commonstrant:

Galileo's earliest work involved the study of the pendulum, inspired by observing a lamp swinging in Pisa cathedral. Following further experiments, he concluded that a pendulum would take the same time to swing back and forth regardless of the amplitude of the swing. This would prove vital in the development of the pendulum clock, which Galileo designed and was constructed after his death by his son.

One of the inventions Galileo is often mistakenly credited with today is the invention of the telescope. This is not true; there had been a number of early prototypes mostly developed in Holland before him, and a Dutch optician called Hans Lippershey applied for a patent on his version in 1608. Galileo did, however, develop his own far superior astronomical telescope from just a description of Lippershey's invention, and quickly employed it to make numerous discoveries. A strong supporter of the Copernican view of planetary motion, Galileo's initial findings published in the *Sidereal Messenger* (1610) provided the first real physical evidence to back up this interpretation. As well as discovering craters and mountains in the moon, sunspots and the moonlike phases of Venus for the first time, he also noted faint, distant stars which supported the Copernican view of a much larger universe than Ptolemy had ever considered. More importantly, he discovered Jupiter had four moons which rotated around it, directly contradicting the still commonly held view, including that of the Church, that all celestial bodies orbited earth, 'the centre of the universe'.

Galileo and Copernicus

Galileo's *Dialogue Concerning the Two Chief World Systems* – Ptolemaic and Copernican in which the Ptolemaic view was ridiculed, attracted the attention of the Catholic Inquisition when it was published in 1632. Threatened with torture, Galileo renounced the Copernican System. His work was placed on the banned 'Index' by the Church where it remained until 1835, and he was subject to house arrest for life. But the tide of scientific revolution Galileo had helped instigate proved too powerful to hold back.

IOANNIS KEPLERI
Mathematici Cælarei
hanc Imaginem

ARGENTORATENSI BIBLIOTHECÆ
Conlecr.

MATTHIAS BERNECCERVS.
Kal. Ianuar. Anno Chr.
M. DC. XXVII.

JOHANNES
KEPLER

CZECHIA
1571–1630

IDEAS AND INVENTIONS
Kepler's Laws of Planetary Motion, Ray Theory of Light

FIELDS
Astronomy, Optics

Opposite: Johannes Kepler.

'THERE IS NO FIGURE LEFT FOR THE ORBIT OF THE PLANET BUT A PERFECT ELLIPSE.'

- Johannes Kepler

The German mathematician, Johannes Kepler, while probably not as well remembered as Copernicus, was one of the key reasons why the Polish astronomer's theories finally became widely accepted. What Copernicus had started in suggesting a heliocentric model of the solar system, i.e. that the planets actually rotated around the sun, Kepler finished in providing the arithmetical and observational proof to support such a thesis.

Kepler himself ultimately owed much of his success to the most famous astronomer of the second half of the sixteenth century, Tycho Brahe, a Dane. Brahe had become aware of Kepler's potential after reading a paper he had written while at university in Tübingen. After Kepler had been forced to leave his post as a mathematics lecturer at Graz in Austria, Brahe invited him to become his assistant in Prague under the patronage of Rudolph II, ruler of the Holy Roman Empire. Kepler took up the post in 1600 and formulated a productive, if somewhat stormy, relationship with Brahe.

One of the reasons the two argued was because Brahe rejected outright the Copernican view of the universe, which Kepler held in such high regard. The Dane had formulated his own alternative and rather obscure view on the rotation of the planets, which never caught on. Although history would subsequently prove Brahe to be wrong, his importance to Kepler,

Kepler and Brahe had a troubled relationship, in part because of Kepler's support for the Copernical model of the universe.

and astronomy in general, was that he was a brilliant observer of the skies and kept excellent records. When Brahe died in 1601, Kepler not only inherited his position of imperial mathematician in Rudolph's court but, crucially, his astronomical notes.

At the time only two new stars visible to the naked eye had been discovered since antiquity. The second was observed by Kepler in 1604.

Using Brahe's records from the previous twenty years, Kepler set about trying to calculate and explain the orbit of Mars. Unfortunately, because he shared Copernicus' view that the planets orbited in perfect circles, the German struggled for the next eight years to produce a satisfactory conclusion. One day, he 'awoke from sleep and saw a new light break' as suddenly he realized that the planets did not rotate in perfect circles at all. They orbited around an ellipse, that is, a (flattened) circle with two 'centres' very close together. At a stroke, this would provide the simple mathematical explanation which had eluded Copernicus and Ptolemy when trying to predict the movement of the planets.

In 1609, Kepler published his findings in *Astronomia Nova* or *New Astronomy*, which crystallized two 'laws' that would have a vital influence on our understanding of the universe. In a later book of 1619, *Harmonices Mundi* or *Harmonies of the World*, he added another important rule. These three together made up 'Kepler's Laws of Planetary

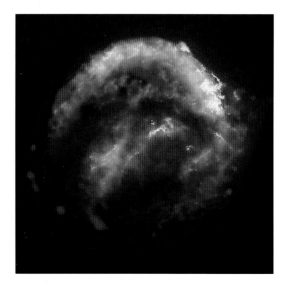

Above: Kepler observed a supernova in 1604 and studied it closely. NASA's space telescopes captured the image 400 years later.

Motion'. The first formalizes his earlier discovery that the planets rotate in elliptical orbits with the sun at one of the centres, or focus points. The second states that all planets 'sweep' or cover equal areas in equal amounts of time regardless of which location of their orbit they are in. This is important because, as the sun is only one of two centres in a planet's orbit, a planet is nearer to the sun at some times than at others, yet it still 'sweeps out' the same area. What this means is that a planet must speed up when it is nearer the sun and slow down when it is further away. Kepler's third law finds that the 'period' (the time it takes to complete one full rotation – a year for the earth for instance) of a planet squared is the same as the distance from the planet to the sun cubed (in astronomical units). This allows distances of planets to be worked out from observing their cycles alone.

As well as providing the credible solution to predicting planetary motion that had previously proved so difficult, Kepler's findings would later act as the stimulus for questions that would lead to Isaac Newton's theory of gravity.

Kepler's last major work was *Tabulae Rudolphinae*, *Ruldolphine Tables* (1627) which were a painstakingly developed series of tables widely used in the next century to help calculate the positions of planets. He also made important discoveries in the field of optics, proposing the ray theory of light.

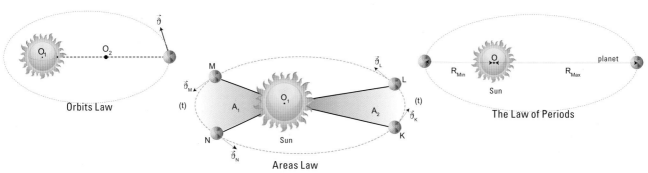

Kepler's three laws of planetary motion.

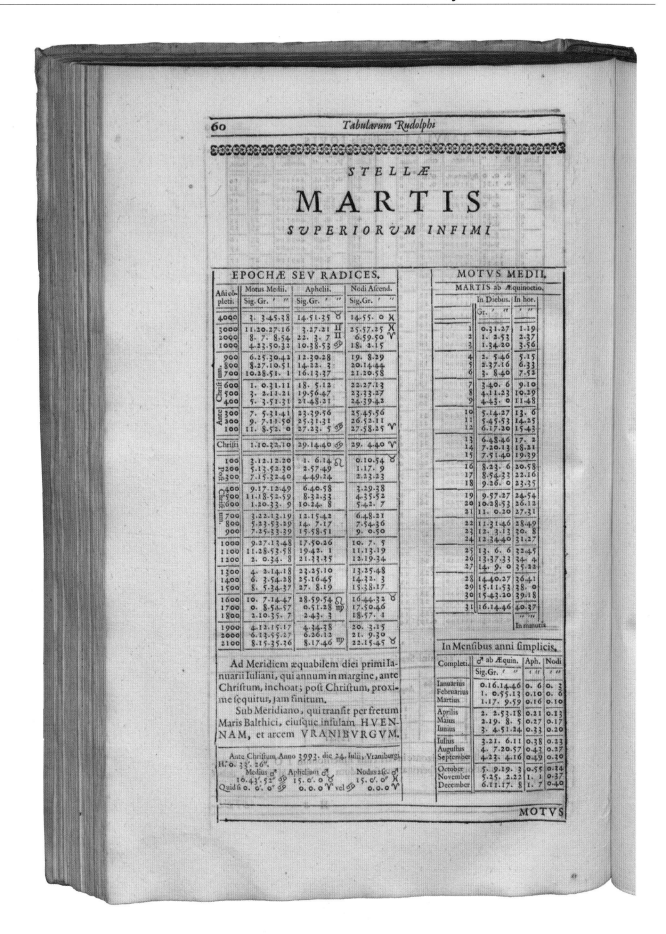

Kepler created the *Rudolphine Tables* to assist with astronomical calculations.

WILLIAM
HARVEY

'THE HEART OF ANIMALS IS THE FOUNDATION OF THEIR LIFE.'

- William Harvey

ENGLAND
1578–1657

IDEAS AND INVENTIONS
Circulatory System, Theory of Reproduction, Anastomosis

FIELDS
Anatomy, Medicine, Biology

Opposite: William Harvey.

If Johannes Kepler thrust astronomy into the modern world by 'completing' the work of Nicolaus Copernicus – who himself had confronted that of Ptolemy – then William Harvey was surely his anatomical equivalent. What Galen had begun and Vesalius had challenged, Harvey credibly launched into the modern arena with perhaps the most significant theory in his field of biology, before or since. What he postulated and convincingly proved was that blood circulated in the body via the heart – itself little more than a biological pump.

Galen had concluded that blood was made in the liver from food which acted as a kind of fuel which the body used up, thereby requiring more food to keep a constant supply. Vesalius, for all his corrections of Galen's work, added little to this theory. So it was left to the Englishman William Harvey, physician to King's James I and later Charles I, to prove his theory of circulation through rigorous and repeated experimentation on the 'royal' stock of animals over two decades.

In the first instance, he had believed the heart could simply not produce the quantities of blood required to support Galen's 'refuelling' theory. To Harvey's mind then, the only sound alternative was that blood was not used up but was recycled around the body. His dissections led him to correctly conclude that the arteries took blood from the heart to the extremities of the body, able to do so because of the heart's pump-like action. The veins, with their series of one-way valves, brought the blood back to the heart again. This rejected Galen's accepted explanation of how the body functioned.

Harvey published his findings in the 720-page *Exercitatio Anatomica de Motu Cordis et Sanguinis in Animalibus* or *Anatomical Exercise on the Motion of the Heart and Blood in Animals* at the Frankfurt Book Fair in 1628. He had, however, been lecturing on his theories of circulation since as early as 1616 but had taken a long time to publish his work. Rather like Copernicus, he was something of a perfectionist, partially explaining why he delayed for so long, but equally he feared a backlash against his theories for challenging Galen head on.

And rightly so. Although he initially received support from some academics, an equal number reacted with outrage and ridiculed his ideas. One of the areas where Harvey's work was weakest, which the author himself acknowledged but had been unable to solve, was that he struggled to offer a proven explanation for how the blood moved from the arteries to the veins. He speculated that the exchange took place through vessels too small for the human eye to see, which was confirmed shortly after his death with the discovery of capillaries by Marcello Malpighi with the recently invented microscope. Harvey, though,

had had no such luxury and even lost patients at his London practice as a result of the criticism directed towards him. By the time of his death, however, he had answered most of his detractors' objections and his conclusions became increasingly accepted, even before Malpighi's final proof.

In 1651, Harvey published another notable work, this time in the area of reproduction. *Exercitationes de* *Generatione Animalium* or *Essays on the Generation of Animals* included conjecture which rejected the 'spontaneous generation' theory of reproduction in mammals which had hitherto persisted. Instead, he suggested the only plausible explanation was that female mammals carried eggs which were somehow spurred into reproduction through interaction with the male's semen. While he did not foresee the egg

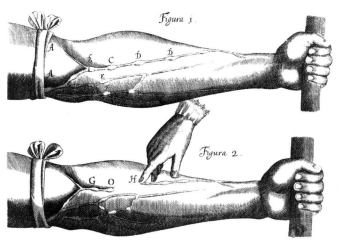

A diagram of experiments used by Harvey to demonstrate the role of valves in the veins.

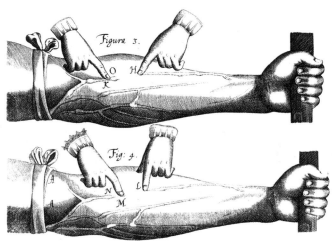

Left: William Harvey demonstrates the circulation of the blood to King Charles I of England.

itself being fertilized in the sense we now understand reproduction, his belief that the egg was at the root of all life was convincing, and gained acceptance long before the observational proof some two centuries later.

Harvey's significance comes not only from his discoveries, but also his methodology. As William Gilbert had begun in physics, and Francis Bacon had subsequently implored in all aspects of life, Harvey was the first to take a rational, modern, scientific approach to his observations in biology, sewing the seeds for a methodolgy that we can accept today. He cast aside the prejudices of his predecessors and only 'induced' conclusions based on the results of experiments which he could repeat identically again and again. It was a model which gained popularity following Harvey's success, and continues to be employed.

BLAISE PASCAL

'MAN IS OBVIOUSLY MADE TO THINK. IT IS HIS WHOLE DIGNITY AND HIS WHOLE MERIT.'

- Blaise Pascal

FRANCE
1623–1662

IDEAS AND INVENTIONS
Digital Calculator, Probability Theory, Pascal's Law, Pascal's Triangle, Pascal's Wager,

FIELDS
Mathematics, Philosophy, Physics

Opposite: Blaise Pascal.

Perhaps one of the lesser noted benefits of being a child prodigy is that if you die at an early age, you have still had sufficient time to fulfil your potential! Blaise Pascal, a Frenchman who passed away at just thirty-nine, was one such example. Although his time on earth was unfortunately cut short by poor health, and his contributions to mathematics and science severely limited by his abandonment of his studies in favour of religious devotion in 1655, he still had a significant influence within both fields of endeavour.

During his twenties Pascal spent a large amount of time undertaking experiments in the field of physics. The most important of these involved air pressure.

An Italian scientist, Evangelista Torricelli (1608–47), had argued that air pressure would decrease at higher altitudes. Pascal set out to prove this by using a mercury barometer. He took initial measurements in Paris and then, at the 1200m-high Puy de Dome in 1646, accompanied by his brother-in-law, he confirmed in no uncertain terms that Torricelli's speculation was true.

More significantly, though, his studies in this area led him to develop Pascal's Principle or Law, which states that pressure applied to liquid in an enclosed space distributes equally in all directions. This became the basic principle from which all hydraulic systems derived, such as those involved in the manufacture of

Pascal's Law shows that a change in pressure at any point in a fluid is transmitted to every other point in that fluid.

Above: Pascal finished his calculating device, the 'Pascaline', in 1644, which could add and subtract numbers.

car brakes, as well as explaining how small devices such as the car jack are able to raise a vehicle. This is because the small force created by moving the jacking handle in a sizeable sweep equates to a large amount of pressure sufficient to move the jack head a few centimetres. Applying the lessons of his studies in a practical way, Pascal went on to invent the syringe and, in 1650, the hydraulic press.

In spite of these developments, however, Pascal is probably better remembered for his work in the area of mathematics. It was here that he showed his genius from an early age. For example, having independently discovered a number of Euclid's theorems for himself by the age of eleven, he went on to master *The Elements*, the great mathematician's definitive text, by twelve. When he was sixteen he published mathematical papers which his older contemporary Descartes at first refused to believe could have been written by one so young. In 1642, still only nineteen, Pascal began work on inventing a mechanical calculating machine which could add and

Above: Pascal is best known for his work in mathematics and in particular his development of the theory of probabilities.

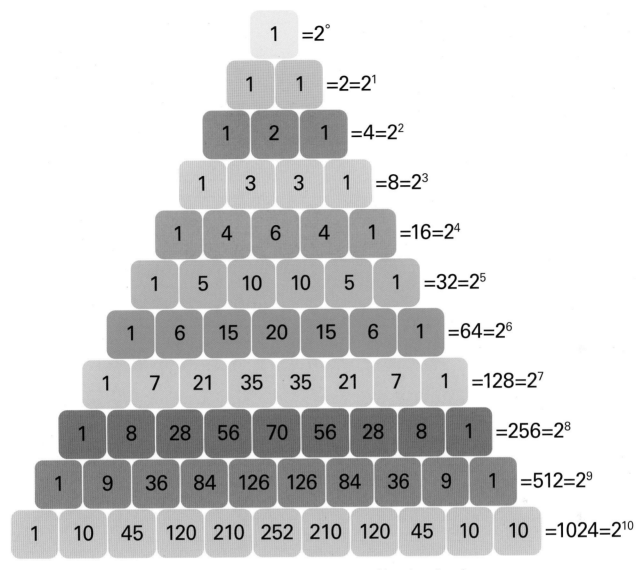

In Pascal's Triangle, each number is the sum of the values above it.

subtract. He had finished what was effectively the first digital calculator by 1644 and presented it to his father to help him in his business affairs.

It was not until later in his short life, around 1654, that Pascal jointly made the mathematical discovery which would have the most impact on future generations. It had begun with a request by an obsessive gambler, the Chevalier de Méré, for assistance in calculating the chance of success in the games he played. Together with Pierre de Fermat, another French mathematician, Pascal developed the theory of probabilities, using his now famous Pascal's Triangle, in the process. As well as its obvious impact upon all parts of the gambling industry, the importance of understanding probability has had subsequent application in areas stretching from statistics to theoretical physics.

The SI unit of pressure – the pascal – and the computer language, Pascal (named in honour of his contribution to computing through his invention of the early calculator), are named after him in recognition of two of his main areas of scientific success.

Seven of the calculating devices that he produced in 1649 survive to this day.

Like many of his contemporaries, Pascal did not separate his science from philosophy, and in his book *Pensées*, he applies his mathematical probability theory to the perennial philosophical problem of the existence of God. In the absence of evidence for or against God's existence, says Pascal, the wise man will choose to believe, since if he is correct he will gain his reward, and if he is incorrect he stands to lose nothing, an interesting, if somewhat cynical argument.

ROBERT BOYLE

GREAT BRITAIN
1627–1691

IDEAS AND INVENTIONS
Boyle's Law, Vacuum Pump, Atomic Theory

FIELDS
Physics, Chemistry

Opposite: Robert Boyle.

'GOD WOULD NOT HAVE MADE THE UNIVERSE AS IT IS UNLESS HE INTENDED FOR US TO UNDERSTAND IT.'

- Robert Boyle

Born in the modern-day Republic of Ireland as the fourteenth child of the richest man in what was then part of Great Britain, Boyle enjoyed all the privileges of an aristocratic education. Schooled at Eton and then privately, he continued his studies as he undertook a long European tour from 1639 to 1644. Eventually he returned to an inherited estate in Dorset, England, largely avoiding the worst excesses of the Civil War. Here he began his scientific studies. In 1656 he moved to Oxford where, with the philosopher John Locke, and the architect Christopher Wren, he formed the Experimental Philosophy Club. He also met Robert Hooke, who became his assistant and the two formed a productive partnership. It was together with Hooke that Boyle began making the discoveries for which he became famous.

Chief amongst these was the expression of what is now known as Boyle's Law (also independently discovered by the French scientist Edme Mariotte) which established a direct relationship between air pressure and volumes of gas. By using mercury to trap some air in the short end of a 'J' shaped test tube, Boyle was able to observe the effect on its volume by adding more mercury. What he found was this: if he doubled the mass of mercury (in effect, doubling the pressure), the volume of the air in the end halved; if he tripled it, the volume of air reduced to a third, and so on. As long as the mass and temperature of the gas were constant, his law concluded that the pressure and volume were inversely proportional.

BOYLE'S LAW

$$P_1V_1 = P_2V_2$$

PRESSURE : P

VOLUME DECREASES
PRESSURE INCREASES

VOLUME INCREASES
PRESSURE DECREASES

VOLUME : V

Boyle's Law showed that the relationship between the pressure and volume of a gas was inversely proportional.

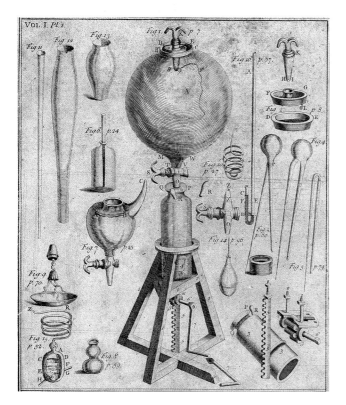

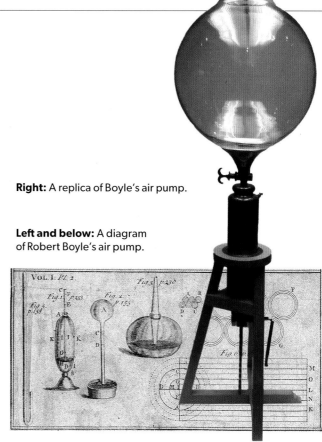

Right: A replica of Boyle's air pump.

Left and below: A diagram of Robert Boyle's air pump.

This experiment was the culmination of a series of other tests involving air and its effects. They had begun shortly after Boyle had moved to Oxford and progressed rapidly when Robert Hooke had constructed an air pump upon his request. The pump was able to create the best man-made vacuum to date and through experiments involving bells, animals and candles, Boyle was able to draw a number of important conclusions. He found that sound could not travel through a vacuum and required air in order to do so. Air was required for respiration and combustion, and not all of the air was used up during breathing and burning processes. In addition, he proved Galileo's proposal that all matter fell at equal speed in a vacuum.

In 1661, Boyle published *The Sceptical Chemist* which criticized the Aristotelian view of a universe composed of only four elements (earth, water, air and fire), plus aether in wider space. The text helped pave the way to our current view of the elements. Although he did not describe elements exactly as we understand them today, he believed that matter consisted at root of 'primitive and simple, or perfectly unmingled bodies' which could combine with other elements to form an infinite number of compounds. This was an extension of his support for early atomic theory,

believing in what he described as tiny 'corpuscles'. In spite of an interpretation which does not entirely correspond with the modern view, his importance was in promoting an area of thought which would influence the later breakthroughs of Antoine Lavoisier (1743–94) and Joseph Priestley (1733–1804) in the development of theories related to chemical elements.

Robert Boyle made a number of contributions to the history of science, but perhaps most significant is his claim to being the man responsible for the establishment of chemistry as a distinct scientific subject in its own right. Like his idol Francis Bacon, he experimented relentlessly, accepting nothing to be true unless he had firm empirical grounds from which to draw his conclusions.

Boyle's other important legacies were the creation of flame tests in the detection of metals, as well as tests for identifying acidity and alkalinity.

Boyle was also a founder member of the Royal Society, the longest running scientific society in the world. But it was his insistence on publishing chemical theories supported by accurate experimental evidence – including, for the first time, details of apparatus and methods used as well as failed experiments – which would have the most impact upon modern chemistry.

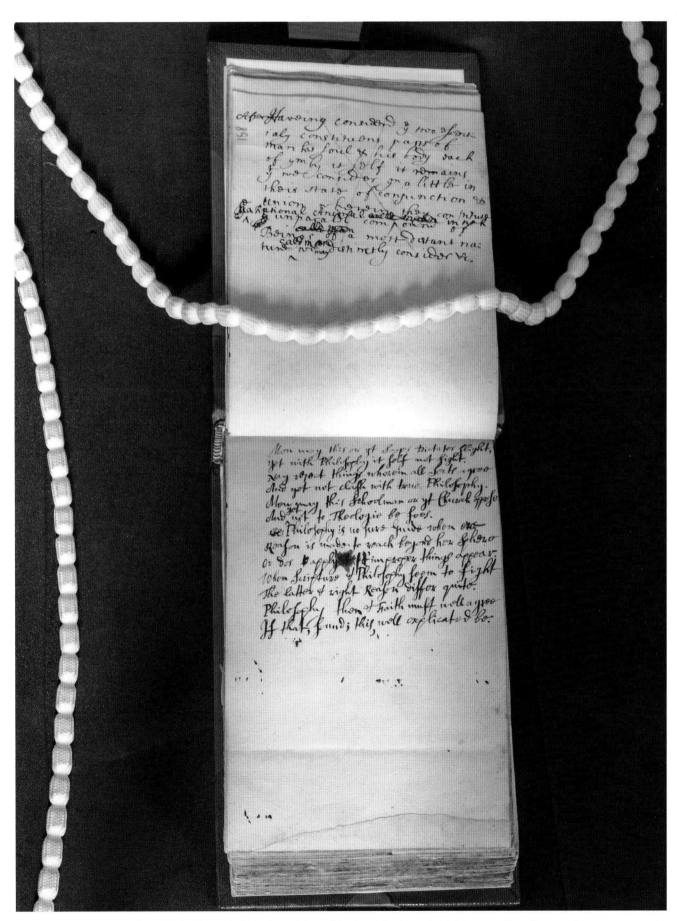

Robert Boyle's notebook from 1690–91, now held in the archives of the Royal Society, of which he was a founding member.

CHRISTIAAN HUYGENS

NETHERLANDS
1629–1695

IDEAS AND INVENTIONS
Discovery of Titan, Pendulum Clock, Refracting Telescope, Wave Theory of Light

FIELDS
Astronomy, Physics

Opposite: Christiaan Huygens.

'HOW MUST OUR WONDER AND ADMIRATION BE INCREASED WHEN WE CONSIDER THE PRODIGIOUS DISTANCE AND MULTITUDE OF THE STARS?'

- Christiaan Huygens

The Dutchman Christiaan Huygens is widely considered to be the second most important physical scientist of the seventeenth century. Son of the distinguished diplomat and scholar Constantjin Huygens, Christiaan was acquainted from an early age with notables such as René Descartes, a friend of the family.

Unfortunately for him, one of his key propositions on the behaviour of light contrasted directly with the first and most important, originally proposed by Isaac Newton. As a result Huygens' theory was largely ignored for over a century. His other achievements in time measurement did impact immediately, however, helping his science to progress in a way it otherwise would have struggled to do.

Isaac Newton articulated a particle theory of light, believing it to be made up of 'corpuscles'. This was a view he summarized in his 1704 text *Opticks* but had held for the preceding decades. He vigorously challenged anyone who tried to contradict this opinion, as both Leibniz and Robert Hooke (1635–1703) – who shared similar views to the Dutchman – were to find out. Huygens believed light actually behaved in a wave-like fashion, in a method which became known as the 'Huygens Construction', which he outlined in his 1690 work *Treatise on Light* (although he had first expressed it in 1678). This opinion much more satisfactorily explained the way light reflected and

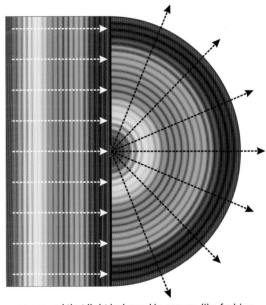

Huygens argued that light behaved in a wave-like fashion.

refracted, and correctly anticipated that in a denser medium light would travel more slowly. Although the modern interpretation is that light can behave in both a particle and wave-like fashion depending on the situation, Huygens' view, when rediscovered and championed by Englishman Thomas Young (1773–1829), in the early nineteenth century, would eventually become the more commonly accepted version. Such was the dominance of Isaac Newton, however, that Huygens' theories were totally ignored for the whole of the eighteenth century, and they still faced fierce resistance in Young's time.

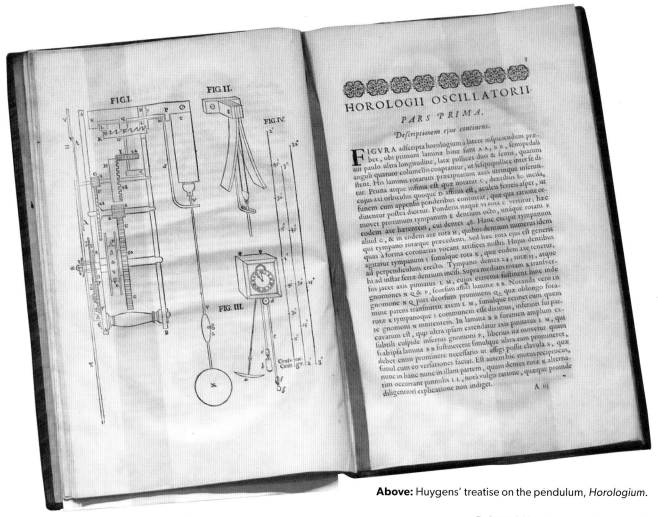

Above: Huygens' treatise on the pendulum, *Horologium*.

Below: A Huygens pendulum clock.

Of much more immediate impact, though, were Huygens' breakthroughs in clock-making. Ever since the time of Galileo (1564–1642), scientists had been aware that a swinging pendulum could keep a regular beat and they had hoped to use this knowledge to create an accurate time-measuring device. They had been unsuccessful. Huygens realized that this was partly because a pendulum mimicking a circle's curve did not maintain a perfectly equal swing and in order to do this it actually needed to follow a 'cycloidal' arc. This discovery set him on the path to designing the first successful pendulum clock. He had it constructed in 1657 and announced his creation to the world in his 1658 book *Horologium* or *The Clock*. The invention was of monumental importance to the progress of physics, for without an accurate method of measuring time the progress of the subject over the following centuries would have been severely hampered.

Huygens backed up his practical findings with

Infrared views of Titan as seen by NASA's Cassini spacecraft. Huygens was the first to discover Titan in 1655.

mathematical explanations describing a pendulum's swing in his 1673 work *Horologium Oscillatorium* or *The Clock Pendulum*. The text included a number of other dynamic explanations and anticipated the first of Newton's motion laws, the 'law of inertia', which states: an object moving in a straight line will continue to move in the same way indefinitely until it meets another force.

Huygens was an associate of Leibniz, whom he supported during his controversial bout with Isaac Newton over the law of gravity. Despite this, and despite Huygens' opinion that Newton's theory of gravity was incomplete without a mechanical explanation, as expressed in the *Principia*, Newton was a staunch admirer of the Dutchman.

As well as being an accomplished physicist, Christiaan Huygens was also a keen astronomer and made some important contributions in this area. He devised a much-improved telescope and used it to make a number of findings, including the discovery of Saturn's biggest moon, Titan, in 1655. In addition, he observed and accurately explained Saturn's ring system.

Huygens' hypothesis that light is a wave was largely ignored at the time as it conflicted with Newton's theory which proposed that light had a particle structure. Both were in fact correct.

Huygens was one of the founding fathers of the French Academy of Sciences in 1666, and was granted a larger pension from that body than anyone else.

ANTONIE
VAN LEEUWENHOEK

*'ON THESE OBSERVATIONS I HAVE SPENT
MORE TIME THAN MANY WILL BELIEVE,
BUT I HAVE DONE THEM WITH JOY.'*

- Antonie van Leeuwenhoek

NETHERLANDS
1632–1723

**IDEAS AND
INVENTIONS**
Microorganisms, protozoa,
spermatozoa, bacteria, blood
cells

FIELDS
Biology

Opposite: Antonie
van Leeuwenhoek.

Who said you had to be a full-time scientist with money and an aristocratic background to make world-changing discoveries? Probably the vast majority of people who lived in the seventeenth century, when most scientists came either from the nobility – possessing the independent wealth to undertake research with no need for a job – or were funded by them through patronage. Not quite so for Anton van Leeuwenhoek, a humble Dutch draper, who, despite little formal education, went on to entertain kings and queens with his remarkable revelations.

Born and remaining all his life in Delft in the Netherlands, van Leeuwenhoek became an apprentice linen-draper at the age of 16 and went on to open his own business in the town around 1654. In 1660 he took on a better-paid position in the town's law courts. This gave him greater means and more spare time to pursue the subject he would bring to impact on history: van Leeuwenhoek had developed a passion for microscopy, and by 1660 was devoting all the spare time he could get to producing lenses with a greater magnification than had ever been made before.

Van Leeuwenhoek kept secret his methods for producing the lenses during his entire ninety years. Even though his finest single, short focal length lenses could enlarge a specimen by up to three hundred times, it is believed he employed an additional technique,

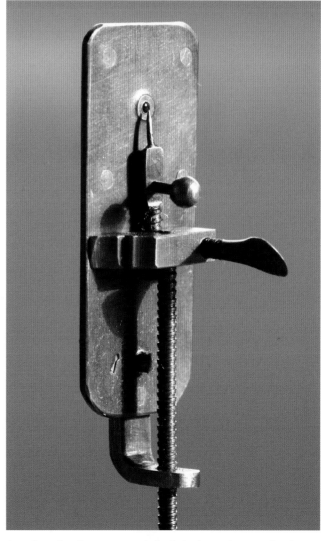

A replica of a microscope made by Antonie van Leeuwenhoek.

perhaps some form of illumination, to view the miniscule 'animalcules' he observed. Another major discovery was protozoa, effectively tiny one-celled plants, which he came across in water specimens in 1674. In more recent scientific studies protozoa would be linked to a number of tropical diseases including, most significantly, malaria and amoebic dysentery. In 1683, and perhaps even more importantly, van Leeuwenhoek observed bacteria for the first time. They were smaller than protozoa and were later linked to diseases such as cholera and tetanus, as well as their treatment.

In between these findings, van Leeuwenhoek discovered spermatozoa. The story goes that in 1677 his contemporary Stephen Hamm brought the microscopist a sample of human semen. Upon examination he discovered the short-lived sperm, reinforcing his opinion of their importance in reproduction by finding similar creatures in the semen of frogs, insects and other animals. He made detailed and exact observations of both fleas and ants, proving that the former were generated from eggs like any other insects, rather than arising spontaneously. He also showed that the eggs and pupae of ants were

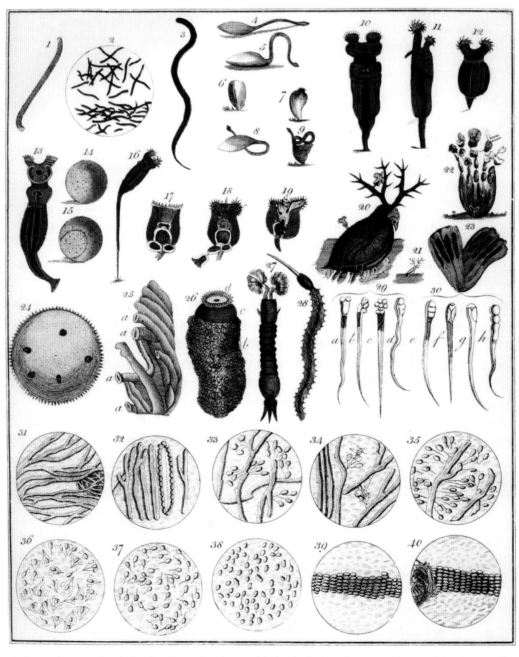

Illustrations made by van Leeuwenhoek showing the various types of 'animalcules' he discovered through his microscopic observations.

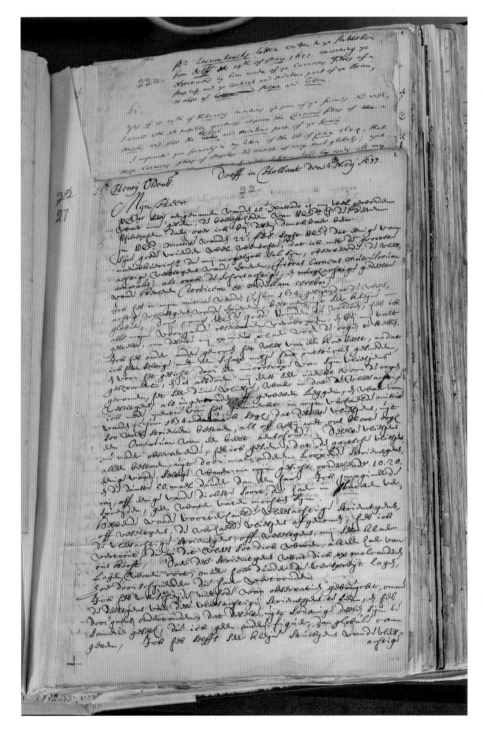

transfer of blood from the arteries to the veins. From ants to shellfish, van Leeuwenhoek also undertook a range of further studies, including observations of animal life.

In keeping with his lack of academic training, van Leeuwenhoek wrote up his findings in Dutch rather than the scholarly Latin, and published little directly. Instead he was introduced to the Royal Society of England via correspondence in 1673. For the rest of his life, van Leeuwenhoek would subsequently write regularly to the Royal Society in Dutch, outlining his latest discoveries, for he knew no English. In all, the society translated and printed some 375 entries in their publication *Philosophical Transactions* before van Leeuwenhoek's death.

Van Leeuwenhoek's letters, and his subsequently assembled collected works, made the part-time scientist world famous and brought many noble visitors to Delft. Amongst those who came to see the animalcules first hand were James II of England and Peter the Great of Russia.

When van Leeuwenhoek died he left behind 247 complete microscopes, nine of which survive to this day. One of his microscopes had a resolution of 2 micrometres.

Examining his own faeces, he observed that 'when of ordinary thickness' there were no protozoa observed, but when 'a bit looser than ordinary', protozoa were observed.

phenomena occurring at two entirely different stages. From this he correctly claimed their existence as evidence of his belief that the commonly held view of 'spontaneous generation' of insects and other small organisms was wrong.

Other important discoveries included the observation of red blood cells in 1684, providing further support for Marcello Malpighi's 1660 work on blood capillaries, which in itself had been so important in reinforcing William Harvey's speculation on the

ISAAC NEWTON

ENGLAND
1642–1727

IDEAS AND INVENTIONS
Gravity, Calculus, Binomial Theory, Laws of Motion, Particle Theory of Light, Theory of Colour

FIELDS
Mathematics, Physics, Optics

Opposite: Isaac Newton.

'IF I SAW FURTHER THAN OTHERS, IT IS BECAUSE I WAS STANDING ON THE SHOULDERS OF GIANTS.'

- Isaac Newton

So many extensive books and articles have been written on the life and impact of Sir Isaac Newton over the last three centuries it is impossible to do his achievements justice in a short entry like this. He is quite simply one of the greatest scientists of all time.

His early years did not necessarily suggest, however, he would end up as such. Born and brought up in the quiet village of Woolsthorpe in Lincolnshire, England, and schooled in the nearby town of Grantham, he was not particularly noted for academic achievements as a child. Even on entry to Trinity College, Cambridge, he did not stand out until, ironically, the University was forced to close during 1665 and 1666 due to the high risk of plague. Newton returned to Woolsthorpe and began two years of remarkable contemplation on the laws of nature and mathematics which would transform the history of human knowledge. Although he published nothing during this period, he formulated and tested many of the scientific principles which would become the basis for his future achievements.

However, it would often be decades before he returned to his earlier discoveries. For example, his ideas on universal gravitation did not re-emerge until he began a controversial correspondence on

UNIVERSAL LAW OF GRAVITATION

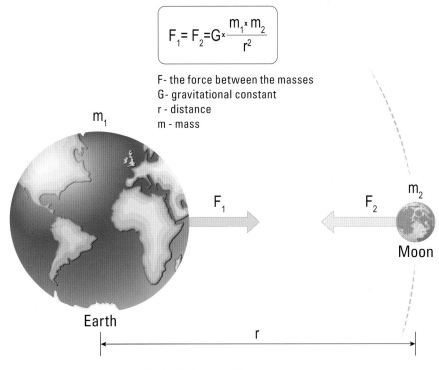

$$F_1 = F_2 = G \times \frac{m_1 \times m_2}{r^2}$$

F- the force between the masses
G- gravitational constant
r - distance
m - mass

m_1

F_1

F_2

m_2

Moon

Earth

r

Newton's Theory of Gravity.

Hooke vs Newton

Robert Hooke was one of the most important scientists of the era. He was the first to describe the universal law that all matter will expand upon heating; he discovered Hooke's Law, which states that the strain, or change in size, placed upon a solid is directly proportional to the stress, or force, applied to it; and he made significant contributions to microbiology through his 1665 work *Micrographia*. As well as this he had a range of inventions to his name, including the vacuum air pump, the reflecting telescope, compound microscope, dial barometer, anemometer, hygrometer, balance spring and more. It was his controversial theory of light – suggesting that it behaved in a wave-like fashion – however, that brought him into conflict with Isaac Newton. Newton vehemently argued against Hooke's theory of light, beginning a bitter feud which would continue for the rest of Hooke's life. Hooke also claimed to have discovered one of the most important theories credited to Newton – gravity. He argued that Newton had plagiarized his ideas from correspondence between the two during 1680, and his letters did suggest some notion of universal gravitation. Nonetheless, Newton's mathematical calculations and endeavours in proving the law give him a much stronger claim.

Left: Robert Hooke.

the subject with Robert Hooke in around 1680. Furthermore, it was not until Edmond Halley challenged Newton in 1684 to find out how planets could have the elliptical orbits described by Johannes Kepler, and Newton replied he already knew, that he fully articulated his law of gravitation. Yet he had begun work on the subject back in the 1660s in Woolsthorpe after famously seeing an apple fall from a tree and wondering if the force which propelled it towards the earth could be applied elsewhere in the universe. After his declaration to Halley, Newton was forced to recalculate his proof, having lost his original jottings, and the result was published in Newton's most famous work *Philosophiae Naturalis Principia Mathematica* (1687). This law of gravitation proposed that all matter attracts other matter with a force related to the combination of their masses, but this attraction is weakened with distance, indeed, in inverse proportion to the square of their distances apart. This universal principle applied just as equally to the relationship between two small particles on earth as it did between the sun and the planets, and Newton was able to use it to explain Kepler's elliptical orbits.

In the same work, Newton built on earlier observations made by Galileo and expressed three laws of motion which have been at the heart of modern physics ever since. The 'law of inertia', states that an object at rest or in motion in a straight line at a constant speed will carry on in the same state until it meets another force. The second stated that a force could change the motion of an object according to the product of its mass and its acceleration, vital in understanding dynamics. The third declares that the force or action with which an object meets another object is met by an equal force or reaction.

Aside from the wide-ranging uses for the laws Newton outlined in the *Principia*, the important point is that all historical speculation of different mechanical principles for the earth from the rest of the cosmos were cast aside in favour of a single, universal system. It was clear that simple mathematical laws could explain a huge range of seemingly disconnected physical facts, providing science with the straightforward explanations it had been seeking since the time of the ancients. Newton's insistence on the use of

Above: Light refracting through a prism. Newton discovered that white light was made up of all the colours of the spectrum.

Right: A replica of Newton's reflecting telescope.

mathematical expression of physical occurrences also underlined the standard for modern physics to follow.

Newton achieved major breakthroughs in other areas too. His proof that white light was made up of all the colours of the spectrum was outlined in his 1672 work *New Theory about Light and Colours*. In *Opticks* (1704), he also articulated his influential (if partially inaccurate) particle or corpuscle theory of light. Another achievement significant to mathematics was his invention of the 'binomial theorem'.

Newton had a practical side too, inventing the reflecting telescope in the 1660s. This new instrument bypassed the focusing problems caused by chromatic aberration in the refracting telescope of the type Galileo had created. During his time as master of the Mint twenty-seven counterfeiters were executed.

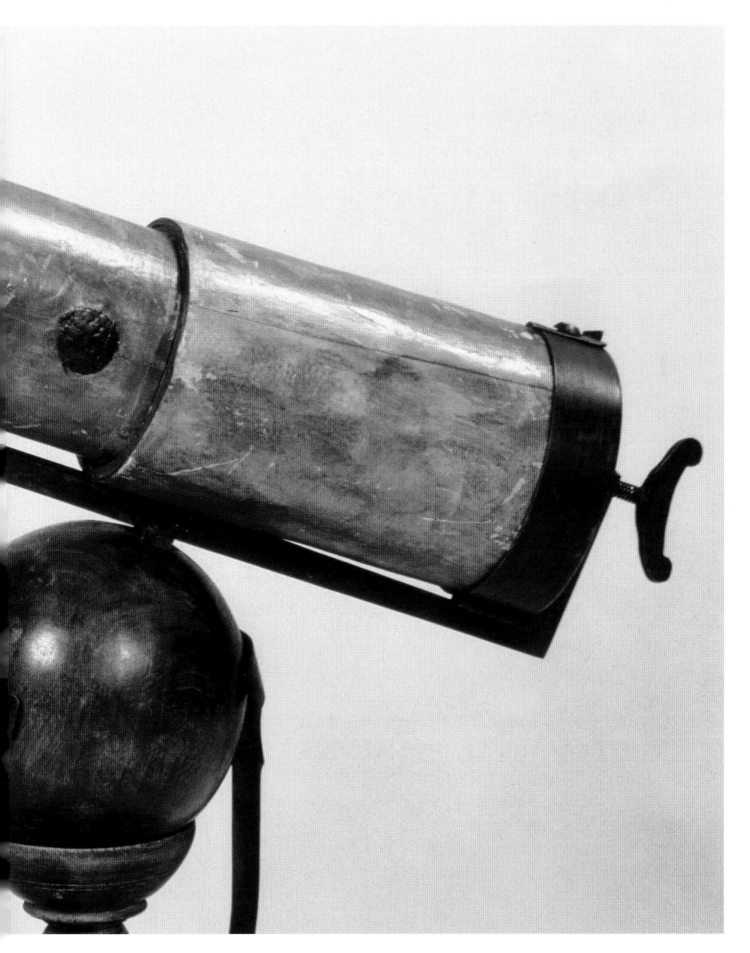

ANTOINE LAVOISIER

FRANCE
1743–1794

IDEAS AND INVENTIONS
Conservation of Matter, Combustion, Oxygen and Hydrogen

FIELDS
Chemistry

Opposite: Antoine Lavoisier.

'WE TRUST TO NOTHING BUT FACTS: THESE ARE PRESENTED TO US BY NATURE, AND CANNOT DECEIVE.'

- Antoine Lavoisier

Despite possible claims to the title by other scientists, the Frenchman Antoine Lavoisier is regarded by most as the true founder of modern chemistry. Although he often undertook similar work to that of Henry Cavendish (1731–1810), Joseph Priestley (1733–1804) and Karl Scheele (1742–86), it was the interpretation of his findings which distinguished Lavoisier. His conclusions led to the restructuring of chemistry into a format which laid the foundations for the modern era, arguably achieving an impact comparable to that of Newton (1642–1727) in physics. For this it would be reasonable to assume the scientist could have expected accolades and awards from his countrymen. Instead, they chopped off his head.

Lavoisier's early studies involved experiments concerning the weight loss or gain in substances when heated. By burning matter such as lead and phosphorus in closed vessels, and accurately weighing them before and after heating, he was able to observe the containers did not gain or lose any mass at all during combustion. This ultimately led to his conclusion of the law of conservation of matter. Lavoisier suggested matter was simply rearranged on heating and nothing was actually added or destroyed overall, hence the equal weight of the vessel before and afterwards. This in itself called into question the 'phlogiston' theory, the commonly held belief all combustible material contained a mysterious element which was released (and lost) on heating.

While the overall weight of the vessel remained the same during Lavoisier's experiments, he made the further interesting discovery that the solids being heated could in fact gain mass. He observed such a reaction, for example, in 1772 when burning phosphorus and sulphur. The only logical conclusion, therefore, was the weight gain of the solid had been caused by some kind of combination with the air trapped in the container. This idea was given further impetus when Lavoisier met Joseph Priestley in Paris in 1774, and the latter explained his discovery of 'dephlogisticated' air. While Priestley failed to realize the impact of this new gas, maintaining a belief in the phlogiston theory, Lavoisier repeated the Englishman's experiments to see if this was the source of the weight gain in some solids during heating. By 1778, he had definitively concluded that not only was Priestley's dephlogisticated air the gas from the atmosphere which was combining with the matter but, moreover, it was actually essential for combustion to take place at all. He renamed it oxygen ('acid producer' in Greek) from the mistaken belief the element was also evident in the make up of all acids. He also noted the existence of the other main component of air, the inert gas nitrogen which he named 'azote'.

The Frenchmen summarized his new order for chemistry in his 1789 book *Traité élémentaire de chimie* or *Elementary Treatise on Chemistry*, sounding

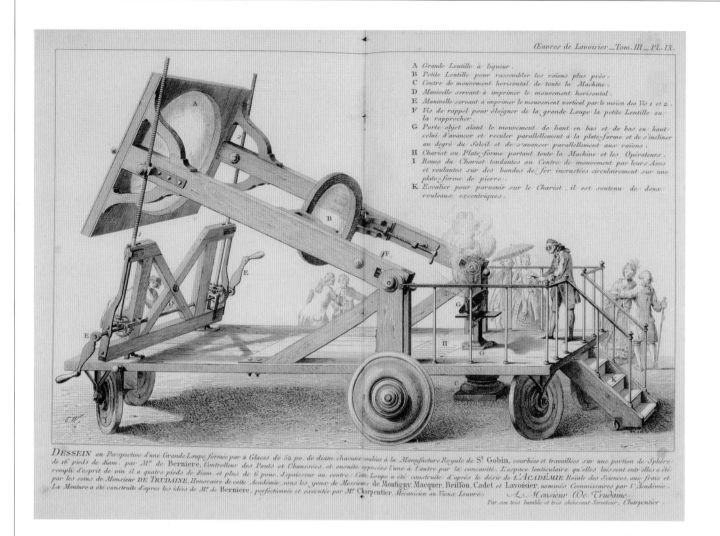

A Grande Lentille à liqueur.
B Petite Lentille pour rassembler les raïons plus près.
C Centre de mouvement horisontal de toute la Machine.
D Manivelle servant à imprimer le mouvement horisontal.
E Manivelle servant à imprimer le mouvement vertical par le moïen des Vis 1 et 2.
F Vis de rappel pour éloigner de la grande Loupe la petite Lentille ou la rapprocher.
G Porte objet aïant le mouvement de haut en bas et de bas en haut, celui d'avancer et reculer parallellement à la plate-forme et de s'incliner au degré du Soleil et de s'avancer parallellement aux raïons.
H Chariot ou Plate-forme portant toute la Machine et les Opérateurs.
I Roues du Chariot tendantes au Centre de mouvement par leurs Axes et roulantes sur des bandes de fer incrustées circulairement sur une plate-forme de pierre.
K Escalier pour parvenir sur le Chariot, il est soutenu de deux rouleaux excentriques.

DESSEIN en Perspective d'une Grande Loupe formée par 2 Glaces de 52 po. de diam. chacune coulées à la Manufacture Royale de St Gobin, courbées et travaillées sur une portion de Sphere de 16 pieds de diam. par Mr de Berniere, Controlleur des Ponts et Chaussées, et ensuite opposées l'une à l'autre par la concavité. L'espace lenticulaire qu'elles laissent entr'elles a été rempli d'esprit de vin il a quatre pieds de diam. et plus de 6 pouc. d'épaisseur au centre. Cette Loupe a été construite d'après le désir de L'ACADÉMIE Roiale des Sciences, aux frais et par les soins de Monsieur DE TRUDAINE, Honoraire de cette Académie, sous les yeux de Messieurs de Montigny, Macquer, Brisson, Cadet et Lavoisier, nommés Commissaires par l'Académie. La Monture a été construite d'après les idées de Mr de Berniere, perfectionnée et exécutée par Mr Charpentier, Mécanicien au Vieux Louvre.
À Monsieur De Trudaine.
Par son très humble et très obéissant Serviteur, Charpentier.

Above: Lavoisier operates his solar furnace, which he built in 1700. It could reach a temperature of 1780°C (3236°F), hot enough to melt platinum.

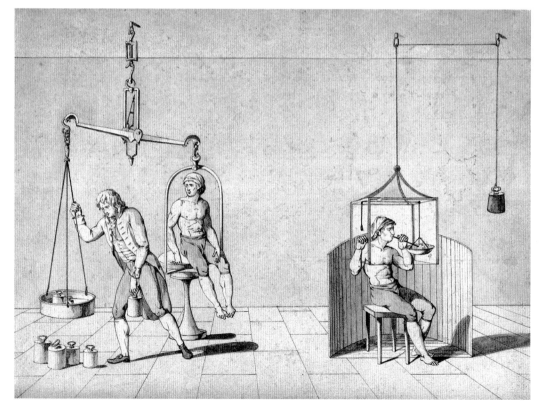

Left: Lavoisier's respiration experiment. It was Lavoisier who determined that oxygen was vital for respiration and at the heart of animal life.

Antoine Lavoisier was tried and executed during the French Revolution.

the death bell for the phlogiston theory and beginning modern chemistry. By this point, Lavoisier had also concluded oxygen was equally vital in the process of respiration, performing a similar kind of role in the body as it did during carbon combustion, and was at the basis of all animal life. In addition, the text included a list of known elements to date, identification work he had begun with a number of other French chemists in the mid-1880s. This founded in turn the naming process of chemical compounds which remains to this day. Lavoisier had already outlined one such combination in 1783, proving water was a combination of hydrogen and oxygen, becoming involved in confusion with James Watt (1736–1819) and Cavendish over who had made the discovery first.

Regarded today as the father of modern chemistry, Lavoisier's name is still used in the title of the modern chemical naming system. Despite the importance of Lavoisier's scientific achievements, they were not enough to save his life in the aftermath of the 1789 French Revolution. Lavoisier was a prominent figure in French public life and, most significantly, ran a tax-collecting firm. Those running such companies were considered to be enemies of the revolution. A prominent member of the revolutionary leadership, Marat, who had earlier attempted to forge a career in science and had had his work criticized by Lavoisier, used this as a pretext to try him. It ultimately led to the guillotine and a tragic end to the life of France's most outstanding scientist.

Firenze Ballagny Lit. Editore.

ALESSANDRO VOLTA

ITALY
1745–1827

IDEAS AND INVENTIONS
Voltage, Batteries

FIELDS
Physics, Chemistry

Opposite: Alessandro Volta.

'EACH METAL HAS A CERTAIN POWER, WHICH IS DIFFERENT FROM METAL TO METAL, OF SETTING THE ELECTRIC FLUID IN MOTION...'

- Alessandro Volta

Whilst the study of electricity had begun to make progress through the work of Benjamin Franklin (1706–90) and others, there was still no reliable way of storing and producing a regular electric current. This had hampered ongoing experimentation in the subject, limiting the scope and usefulness of investigations. One scientist fascinated by electricity and determined to overcome this hurdle was Alessandro Volta.

The Italian aristocrat was born in Como, Lombardy, into a family where most of the male line had entered into the priesthood. Science was clearly Volta's calling, however, and in 1774 he became first teacher and shortly afterwards professor of physics at the Royal School in his hometown. Within a year, he had developed his first major breakthrough in the field of electricity with his invention of the 'electrophorus', used in the production and storage of static electricity.

This device brought Volta recognition within his field and in 1779 he was offered the chair of physics at Pavia University, a post he accepted and would go on to hold for the next quarter of a century. Here he continued his electrical investigations, becoming particularly absorbed in the work of Luigi Galvani (1737–98) during the 1780s. Volta's countryman had made a strange discovery during dissection work. He had found that simply by touching a dead frog's legs with two different metal implements, the muscles in the frog's legs would twitch. Through various other experiments, Galvani wrongly concluded it was the animal tissue which was somehow storing the electricity, releasing the substance when touched by the metals.

Volta, however, was not convinced the animal muscle was the important factor. He set about recreating Galvani's experiments and concluded, controversially at the time, the different metals were the important factor in the production of the current. Indeed, Volta and Galvani had been friends before the former began criticizing the deductions his peer had made concerning the importance of the animal tissue. To make matters worse, it was Galvani himself who had sent Volta his papers on the subject for Volta's review and, he hoped, support. Instead, a bitter dispute broke out concerning whose analysis was correct. Although Galvani would not live long enough to see Volta's ultimate rebuttal of his work, the argument was already swinging in the latter's favour by the time of Galvani's death, and he ended his days a disillusioned man.

To back up his theory, Volta had begun putting together different combinations of metals to see if they produced any current, even going as far as to use his tongue as an indicator of current strength from the shock they produced! It was, in fact, an important test because he deduced the saliva from his tongue

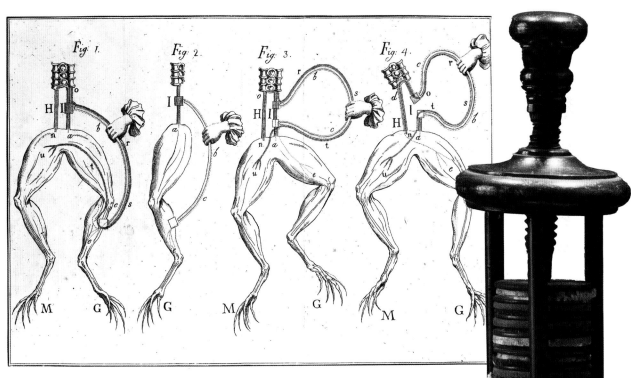

was a factor in aiding the flow of the electric current. Consequently, Volta set about producing a 'wet' battery of fluid and metals. His decisive solution came in 1800, with the 'Voltaic pile', a stack of alternating silver and zinc disks interspersed with brine-soaked cardboard layers. By attaching a copper wire to the ends of this device and closing the circuit, Volta found it produced a regular, flowing electric current. He had created the first battery.

The invention radically improved the study of electricity, facilitating further breakthroughs in the subject by other scientists such as William Nicholson and Humphrey Davy (1778–1829), who made discoveries using electrolysis, and later aided the work of Michael Faraday (1791–1867). Napoleon, who at that time controlled the territory in which Volta lived, invited the scientist to demonstrate his invention in Paris in 1801. He was so impressed that he made Volta a count, and later a senator, of Lombardy, and awarded him the Legion of Honour medal.

The volt, the SI unit of electric potential, is named after the Italian. A volt is defined as the difference of potential between two points on a conductor carrying one ampere current when the power dissipated between the points is one watt.

Volta was also the first to isolate methane gas, an achievement made in 1778.

Above: Luigi Galvani believed that the twitching frog legs in his experiments were a result of 'animal electricity'. Alessandro Volta correctly ascertained that the reaction was not due to the animal, but a result of the metals used.

Right: A Voltaic pile from 1801. These early devices were the first electric batteries.

Alessandro Volta demonstrates his Voltaic pile to Napoleon. He received a Legion of Honour Medal for his efforts.

JOHN DALTON

ENGLAND
1766–1844

IDEAS AND INVENTIONS
Atomic Theory, Law of Multiple Proportions, Law of Partial Pressures

FIELDS
Physics, Chemistry

Opposite: John Dalton.

'ALL THE CHANGES WE CAN PRODUCE, CONSIST IN SEPARATING PARTICLES THAT ARE IN A STATE OF COHESION OR COMBINATION, AND JOINING THOSE THAT WERE PREVIOUSLY AT A DISTANCE.'

- John Dalton

For much of his life, the primary interest of the English Quaker, John Dalton, was the weather. Living in the notoriously wet county of Cumbria, he maintained a daily diary of meteorological occurrences from 1787 until his death, recording in total some 200,000 entries. Yet, it was his development of atomic theory for which he is most remembered.

It was around the turn of the nineteenth century that Dalton started to formulate his theory. He had been undertaking experiments with gases, in particular on how soluble they were in water. A teacher by trade, who only practised science in his spare time, he had expected different gases would dissolve in water in the same way, but this was not the case. In trying to explain why, he speculated that perhaps the gases were composed of distinctly different 'atoms', or indivisible particles, which each had different masses. Of course, the idea of an atomic explanation of matter was not new, going way back to Democritus of Abdera (c. 460–370 BCE) in ancient Greece, but now Dalton had the discoveries of recent science to reinforce his theory. On further examination of his thesis, he realized that not only would it explain the different solubility of gases in water, but would also account for the 'conservation

The ancient Greek philosopher Democritus was the first to suggest that there were individual atoms which made up all matter.

83

of mass' observed during chemical reactions as well as the combinations into which elements apparently entered when forming compounds (because the atoms were simply 'rearranging' themselves and not being created or destroyed).

Dalton publicly outlined his support for this atomic theory in a lecture in 1803, although its complete explanation had to wait until his book of 1808 entitled *A New System of Chemical Philosophy*. Here, he summarized his beliefs based on key principles such as: atoms of the same element are identical; distinct elements have distinct atoms; atoms are neither created nor destroyed; everything is made up of atoms; a chemical change is simply the reshuffling of atoms; and compounds are made up of atoms from the relevant elements. In the same book he published a table of known atoms and their weights, although some of these were slightly wrong due to the crudeness of Dalton's equipment, based on hydrogen having a mass of one. It was a basic framework for subsequent atomic tables, which are today based on carbon (having a mass of 12), rather than hydrogen. Dalton also wrongly assumed elements would combine in one-to-one ratios (for example, water being HO not H_2O) as a base principle, only converting into 'multiple proportions' (for example, from carbon monoxide, CO, to carbon dioxide, CO_2) under certain conditions. Although the debate over the validity of Dalton's thesis would continue for decades, the foundations for the study of modern atomic theory had been laid and with ongoing refinement were gradually accepted.

Prior to atomic theory, Dalton had also made a number of other important discoveries and

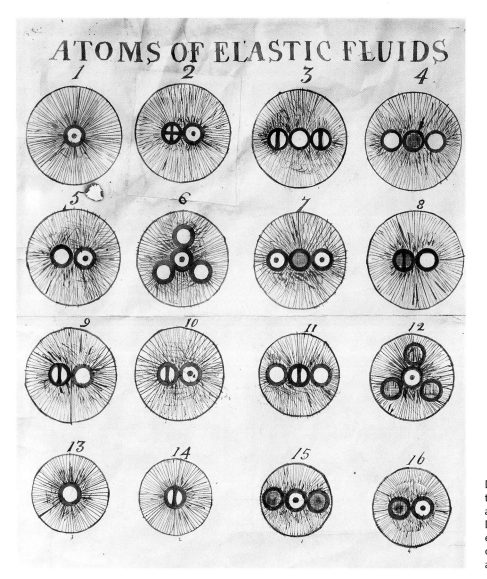

Diagrams of the structure of atoms by John Dalton. Dalton established the core tenets of atomic theory.

The aurora borealis fascinated John Dalton and he theorized that they had a magnetic cause.

observations in the course of his work. These included his 'law of partial pressures' of 1801, which stated that a blend of gases exerts pressure which is equivalent to the total of all the pressures each gas would wield if they were alone in the same volume as the entire mixture.

Dalton also explained that air was a blend of independent gases, not a compound. He was the first to publish the law later credited to and named after Jacques-Alexandre-César Charles (1746–1823). Although the Frenchman had been the first to articulate the law concerning the equal expansion of all gases when raised in equal increments of temperature, Dalton had discovered it independently and had been the first to print.

Dalton also discovered the 'dew point' and that the behaviour of water vapour is consistent with that of other gases, and hypothesized on the causes of the aurora borealis, the mysterious Northern Lights. His further meteorological observations included confirmation of the cause of rain being due to a fall in temperature not pressure.

John Dalton began teaching at his local school at the age of 12. Two years later he and his elder brother purchased a school where they taught roughly 60 children.

His paper on colour blindness, which both he and his brother suffered from, and which was known as daltonism for a long while, was the first to be published on the condition. Dalton is also largely responsible for transforming meteorology from being an imprecise art based on folklore to a real science; how much more precise it is nowadays is perhaps debatable!

MICHAEL FARADAY

ENGLAND
1791–1867

IDEAS AND INVENTIONS
Electric Motor, Electrolysis, Faraday Effect, Law of Induction

FIELDS
Physics, Chemistry

Opposite: Michael Faraday.

'NOTHING IS TOO WONDERFUL TO BE TRUE, IF IT BE CONSISTENT WITH THE LAWS OF NATURE.'

- Michael Faraday

Michael Faraday is regarded as one of the greatest experimental scientists of all time. Even Albert Einstein (1879–1955) considered him to be one of the most important influences in the history of physical science. Yet the man whose discoveries and inventions, amongst them the electric motor, electric generator and the transformer, were to have such a profound impact on modern life, might not have entered the scientific arena at all but for certain fortuitous events in his youth. The first was his apprenticeship at a bookbinder's when he was thirteen. Here his interest in science and in particular electricity was stimulated upon reading pages from the books he was tasked to bind. The second fortunate incident was his appointment as assistant to the renowned chemist Sir Humphrey Davy (1778–1829), who had remembered the young Faraday attending his lectures. The temporary post soon turned permanent and shortly afterwards Davy took Faraday with him on a grand European tour which gave the young man the rare opportunity to meet and learn from many of the leading physicists and chemists of the day.

A magnetic sparking coil made by Michael Faraday. A precursor to the dynamo, it showed how electricity could be generated from a magnetic field.

Michael Faraday gives a Christmas lecture to the public in 1855.

Much of Faraday's early work as a scientist in the 1820s was not in physics, the area which ultimately led to his breakthrough inventions, but in chemistry. In 1823, he became the first person to liquify chlorine, albeit accidentally, while he was conducting another experiment. He quickly deduced how the new form of chlorine had been obtained and applied the process, which made use of pressure and cooling, to other gases. By employing his talent as an outstanding analyst of his own chemical experiments, he also went on to discover benzene in 1825.

Yet it is physical science, in particular his work involving electricity, for which Faraday is best remembered today. As early as 1821, he was able to create the first electric motor after discovering electromagnetic rotation. He had developed Hans Christian Oersted's (1777–1851) 1820 discovery that electric current could deflect a magnetic compass needle. Faraday's experiment proved that a wire carrying an electric current would rotate around a fixed magnet and that conversely,

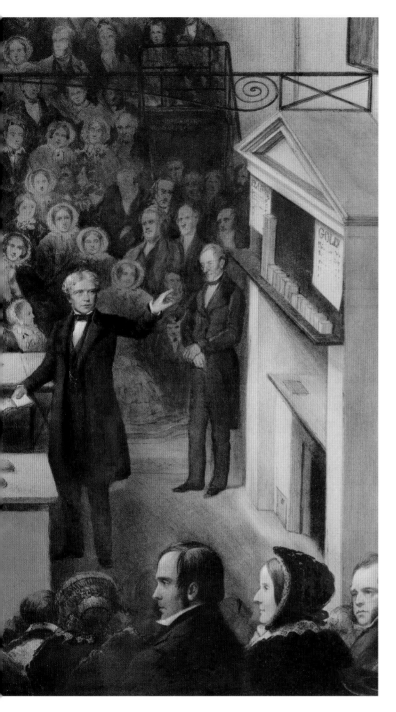

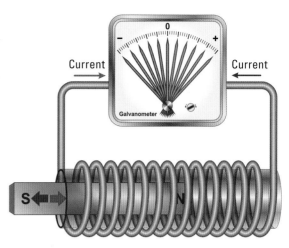

Electromagnetic induction.

Joseph Henry) and even the dynamo: inventions which can truly be claimed to have changed the world!

The reason Faraday was able to make such advances was because from early in his career he had rejected the concept of electricity as a 'fluid', an idea that had been accepted up until that time, and instead visualized its 'fields' with lines of force at their edges. He believed that magnetism was also induced by fields of force and that it could interrelate with electricity because the respective fields cut across each other. Proving this to be true by producing an electric current via magnetism, Faraday had discovered electromagnetic induction. He was encouraged by this and went on to explore the idea that all natural forces were somehow 'united'. He then focused on how light and gravity were related to electromagnetism. This led to the discovery of the 'Faraday effect' in 1845 which proved that polarized light could be affected by a magnet. James Clerk Maxwell proved that light was indeed a form of electromagnetic radiation, and eventually provided the mathematical expression for Faraday's law of induction.

Faraday's fascination with electricity and his background in chemistry both found a natural expression in electrolysis, in which he was also to perform ground-breaking work. In 1833 he was the first to state the basic laws of electrolysis, namely that: (1) during electrolysis the amount of substance produced at an electrode is proportional to the quantity of electricity used and (2) the quantities of different substances left on the cathode or anode by the same amount of electricity are proportional to their equivalent weights.

the magnet would revolve around the wire if the experiment were reversed. From this work, Faraday became convinced that electricity could be produced by some kind of magnetic movement alone but it took ten further years before he successfully proved his hypothesis. In 1831, by rotating a copper disk between the poles of a magnet, Faraday was able to produce a steady electric current. This discovery allowed him to go on to produce electrical generators, the transformer (also invented independently at around the same time by an American,

CHARLES
DARWIN

ENGLAND
1809–1882

**IDEAS AND
INVENTIONS**
Evolution, Natural Selection

FIELDS
Natural History, Biology,
Geology

Opposite: Charles Darwin.

*'MAN, WITH ALL HIS NOBLE QUALITIES,
STILL BEARS THE INDELIBLE STAMP OF
HIS LOWLY ORIGIN.'*

- Charles Darwin

Onboard the HMS *Beagle* during its second expedition from 1831–36 was the young naturalist Charles Darwin. His observations from this time provided the spark for the development of his theory of evolution.

The spark for Darwin's accomplishments was ignited with the 1831 HMS *Beagle* expedition, which was to chart coastlines in the South Americas and other areas of the Pacific. Darwin, supposedly studying religion at that time, had become increasingly absorbed with natural history and persuaded the Professor of Botany, John Henslow, to put him forward for the post of unpaid naturalist on the *Beagle*'s voyage. He thereby abandoned his university studies. His father, and initially the vessel's Captain FitzRoy, resisted, but he eventually persuaded them to let him take part in the five-year expedition.

The Galapagos Islands are especially isolated and provided a unique series of comparisons, with distinct variations between each of them.

During the journey, Darwin made many geological and biological observations, but it was his time spent around the Galapagos Islands which would end up having the most significant impact on him. The ten islands are relatively isolated, even from each other, and as such act as a series of distinct observatories through which Darwin could draw comparisons. He noted that the islands shared many species of flora and fauna in common, but that each land mass often displayed distinct variations within the same group of organisms. For example, he famously noted fourteen different types of finch across the islands, notably with different shaped beaks. In each instance the particular beak seemed to best suit the capture of that bird's prevalent food source, whether it be seeds, insects or fish.

Over the ensuing years, and upon his return to England, Darwin pondered on the reasons for the variations in the finches and other plants and animals. He soon surmised that the birds had descended from a single parent species, rather than each springing up independently and thus acknowledged the idea of evolution, a concept which had existed for some time but was not widely accepted. Darwin began looking for an explanation for this evolution. One text which

had a particular impact on him was Thomas Malthus' 1798 work *An Essay on the Principle of Population* which Darwin read in 1838. Malthus had been concerned with overpopulation resulting in famine, and the possible competition for food which could ensue. Darwin immediately saw that this could also be applied to the animal world too, where only those best adapted to food collection in their environments would survive. Those that could not compete would die out and the characteristics of the successful

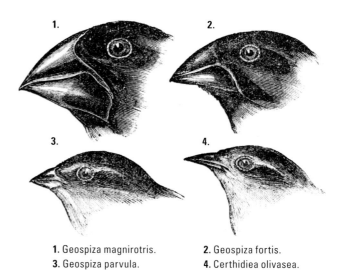

1. Geospiza magnirotris.
2. Geospiza fortis.
3. Geospiza parvula.
4. Certhidiea olivasea.

Illustrations of the Galapagos finches' beaks.

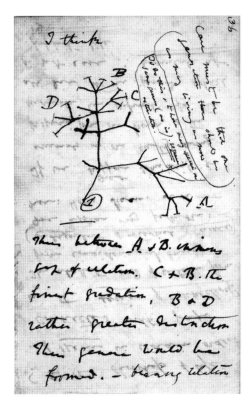

Darwin's first sketch of an evolutionary tree, from 1837.

The predicted outcry ensued and a fierce debate followed, but Darwin already had a number of friends, particularly Thomas Huxley, known as 'Darwin's Bulldog', who would vigorously defend his ideas. This left Darwin free to follow through further implications of his hypothesis in other works, including the 1871 text *The Descent of Man*, which articulated the idea of the evolution of the human race from other creatures.

Darwin's ideas took a long time to become generally accepted (even today they are not embraced by everyone), challenging as they did all previous conceptions of what it meant to be human. As has been the case with so many scientists, he encountered paticularly fierce opposition from the Church, whose members preferred the safety of a sacred text to the uncertainties of observation and experiment.

The idea of evolution through natural selection is, however, at the heart of modern biology. The man who disappointed his father for lack of academic interest had eventually gone on to turn an entire branch of academia on its head.

animals, which may have occurred in the first place by chance, would be passed on to future generations. As environments changed and animals moved about, success criteria would change, gradually resulting in variations within species, as had happened with the finches. Ultimately new species would also be created.

Unfortunately, such a hypothesis would challenge the commonly held view of man as the lord of the earth, specifically created and placed upon the planet in God's image, as described in the Bible. Darwin was implicitly suggesting that man had evolved by chance over thousands of years. He correctly anticipated uproar and resistance to his ideas, particularly from religious leaders. Consequently, he kept his theories dark for twenty years while he gathered additional evidence to back up his case.

He finally published in 1858. He did this jointly with Alfred Russell Wallace (1823–1913), whose independent ideas were remarkably similar to Darwin's. They agreed to a joint public declaration of their hypotheses by submission of a paper to the Linnean Society. Darwin followed this up with a more detailed account in 1859 containing evidence he had collected over the previous decades called *On the Origin of Species by Means of Natural Selection*.

Darwin's theory provoked a furore. This cartoon from 1871 caricatured Darwin and his ideas.

LOUIS PASTEUR

FRANCE
1822–1895

IDEAS AND INVENTIONS
Pasteurization, Anthrax Vaccine, Rabies Vaccine

FIELDS
Medicine, Biology, Chemistry

Opposite: Louis Pasteur.

'SCIENCE KNOWS NO COUNTRY, BECAUSE KNOWLEDGE BELONGS TO HUMANITY, AND IS THE TORCH WHICH ILLUMINATES THE WORLD.'

- Louis Pasteur

Louis Pasteur's name is best remembered for his development of the process of 'pasteurization'. Though Pasteur was a chemist his most significant breakthroughs were in medicine. Indeed, he is considered by many to be the most important figure in nineteenth century medical research. Much of this reputation hinges on his development of a vaccine against rabies. After Edward Jenner's (1749–1823) breakthrough of a vaccine against smallpox made at the end of the previous century, little more had been done to take advantage of the potential of this treatment against other diseases. In 1880, however, Pasteur was to recognize and manipulate a chance occurrence that he noticed in his laboratory to finally systematize a scientific approach to the development of vaccines.

Some chicken cholera bacteria had accidentally been left alone for a long period. Pasteur noticed that when he injected this into chickens they did not develop, or only suffered a mild form of, the disease normally associated with the bacteria. When he later injected the same chickens with fresh bacteria, they survived, while others which had not received the earlier treatment quickly died. Pasteur drew parallels between this result and the work of Jenner and set about deliberately applying the approach to other diseases.

By 1882 he had successfully produced a vaccination against anthrax, a disease which seldom affected humans, but which could devastate stocks of sheep and cattle. By 1885 he had developed a vaccine, extracted from the spines of infected rabbits, to successfully treat animals for rabies.

Pasteur's apprehension at performing a trial on humans was cast aside when a nine-year old called Joseph Meister was brought to him. The boy had been bitten several times by a rabid dog. Pasteur injected him with the new vaccine and the boy survived. Word of the success spread and the following year over 2,500 infected patients received the same treatment. As a result, fatalities dropped to less than 1 per cent. As well as the immediate benefit and fame that Pasteur's development brought him, it also prompted a rush by other scientists to begin searching for new vaccines for other diseases. Several more successes were heralded by the end of the century.

Prior to this, Pasteur had helped limit the spread of tuberculosis and typhoid through the application of his pasteurization process. This was developed during his studies on the fermentation of milk and alcohol. Through microscopic examination and other experiments he conclusively countered the prevailing argument of the day which held that it was merely a chemical process. Pasteur proved that microorganisms

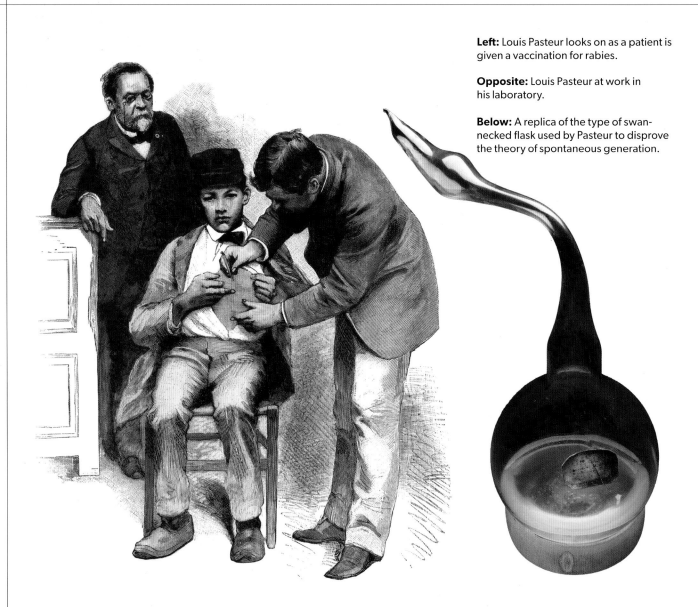

Left: Louis Pasteur looks on as a patient is given a vaccination for rabies.

Opposite: Louis Pasteur at work in his laboratory.

Below: A replica of the type of swan-necked flask used by Pasteur to disprove the theory of spontaneous generation.

were essential for fermentation to take place. He also found that potentially dangerous microbes existing in milk, such as those which caused tuberculosis and typhoid, could be destroyed by heating the liquid for about thirty minutes at 63°C. This is now known as pasteurization, still used to treat milk.

During the same period of work, Pasteur also conclusively disproved 'spontaneous generation' theories, which had persisted for centuries. He demonstrated that sterilized fluid not exposed to microbes in the air would remain uncontaminated, whereas the moment the liquid was put into contact with them it became spoiled.

In addition, from 1865 he greatly aided the French silk industry. By analysing diseases which decimated silkworms he eventually provided successful recommendations for their prevention. Pasteur undertook important work early in his career on the discovery of asymmetrical molecules in compounds which did much for the later development of structural chemistry.

Pasteur used a similar process to pasteurization to improve the success of fermentation in the wine and beer-making industries.

By the time of his death, Pasteur was world famous and tributes poured in. Perhaps the most dramatic gesture of all, however, came almost a half-century later. The nine-year-old boy, Joseph Meister, whom Pasteur had saved from rabies, went on to become caretaker at the Pasteur Institute (founded in 1888) where the scientist was buried. In 1940 the Nazis arrived in Paris and ordered Meister to open Pasteur's tomb in order to examine it. Meister chose to kill himself rather than comply with the violation.

JOHANN GREGOR MENDEL

AUSTRIA
1822–1884

IDEAS AND INVENTIONS
Heritability, Law of Independent Assortment

FIELDS
Biology, Genetics

Opposite: Johann Gregor Mendel.

'WHEN TWO PLANTS, CONSTANTLY DIFFERENT IN ONE OR SEVERAL TRAITS, ARE CROSSED, THE TRAITS THEY HAVE IN COMMON ARE TRANSMITTED UNCHANGED TO THE HYBRIDS AND THEIR PROGENY.'

- Johann Gregor Mendel

The work of the Austrian monk, Johann Gregor Mendel, would be at the heart of the future development of biology, and founded a new branch of science in its own right. During his lifetime and for some time afterwards, however, his efforts were largely ignored. Only when others started making similar discoveries in heredity and began looking for related studies was it realized that Mendel had got there decades before them.

The later impact of Mendel's findings, which effectively act as the starting point for the modern science of genetics, was even more remarkable given the limited amount of his 'career' actually spent researching them. Up to 1856 his time was spent on religious duties or training at his monastery, or in trying to improve his limited early education sufficiently to allow him to pass his teaching examination. Ironically he never succeeded in attaining the qualification, in part due to his lack of success in biology! From 1868 he became abbot of his monastery, located in the modern day Czech Republic, and had to give up the majority of his scientific research. This meant he completed only twelve years of active experimentation.

Even the arena for the breakthrough was an unusual one. Mendel's laboratory was the monastery's garden, his subject the humble pea-plant. The monk had been fascinated by what caused the different characteristics of these plants to occur, such as blossom colour, seed colour and height. He decided to undertake a systematic study of when these features occurred in descendent generations. He set about cross-fertilizing plants with different characteristics and recording the results.

Common assumption at the time was that when two alternate features were combined, an averaging of these features would occur. So, for example, a tall plant and a short one would result in a medium height offspring. The statistical results Mendel collaborated, however, proved something entirely different. Across a series of generations of descendants, plants did not average out to a medium, but instead inherited the original features (for example, either tallness or shortness) in a ratio of 3:1, according to the 'dominant' trait (in the example of height, this was the tall characteristic). He explained this by assuming each parent carried two possibilities for any given trait, for example a tall

St Thomas' Abbey in Brno, in what is now the Czech Republic, where Mendel conducted his experiments.

'gene' (as we now know it) and a short one for height, or a dark gene and light one for seed colour, or gene 'A' and gene 'B' for 'X' trait. Only one gene from each parent would carry into the offspring (now described as Mendel's law of segregation), however, giving four possible combinations: AA, AB, BA and BB. The 3:1 ratio would be achieved because the 'dominant' gene would feature whenever it were present. So if 'A' were the dominant factor, it would occur three times in four, with the 'B' scenario only occurring when a BB result was obtained. He also noted the different pairs of genes making up the characteristics of the pea plant, such as the two causing height, the two causing seed colour, and so on, when crossed occurred in all possible mathematical combinations, independently of each other. This is now described as Mendel's law of independent assortment, and offered him a simple statistical model for predicting the variety of descendants, backed up by ongoing experimental proof.

Mendel first articulated his results in 1865 and published them in an article of 1866 entitled *Experiments with Plant Hybrids*. He was frustrated that the conclusions were largely ignored in his lifetime and it was only when three other scientists, Hugo de Vries (1848–1935), Karl Erich Correns (1864–1933) and Erich Tschermak von Seysenegg

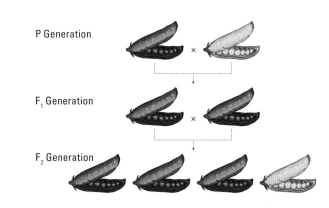

P Generation

F_1 Generation

F_2 Generation

Mendel's experiments with peas showed how a dominant gene would appear at a ratio of 3:1.

Hugo de Vries was another scientist who independently discovered the principle of heredity, leading to the rediscovery of Mendel's ideas in 1900.

(1871–1962) independently came across similar experimental evidence in 1900 that Mendel's work was rediscovered. Its importance in explaining principles of heredity across all sorts of life forms (although with refinement in some areas) was soon realized, and it was later used to underpin Darwin's argument for natural selection too. The science of what is now known as genetics gradually evolved, and Mendel's position as its, albeit unwitting, founder became cemented in history.

Although he did not gain any recognition for his work on heredity during his lifetime, he was well respected and liked by his fellow monks and townspeople. Nowadays, Mendel is regarded as the father of the study of genetics.

Genetics Revised

The 'rediscovery' in 1900 of the laws of inheritance first observed by Johann Gregor Mendel (1822–84) excited many biologists who believed they had found an explanation for hereditary traits, and quite possibly the mechanism to underpin Darwin's theories. The American scientist Thomas Hunt Morgan refined Mendel's ideas, by suggesting that groups of genes had to be present on a single chromosome. He discovered that certain genetic traits did not occur independently of each other but were passed on in groups. Mendel's principle of independent assortment did apply, but only to genes found on different chromosomes. For those on the same chromosome, linked traits would be passed on, usually a sex-related factor with other specific features. The results of his work convinced Morgan that genes were arranged on chromosomes in a linear manner and could actually be 'mapped'. The nearer on the chromosome the genes were located to each other, the less likely the linkages were to be broken. Thus, by measuring the occurrence of breakages, he could work out the position of the genes along the chromosome. Consequently, in 1911, he produced his first chromosome map, showing the position of five genes which were linked to gender characteristics.

Thomas Hunt Morgan.

JAMES CLERK
MAXWELL

'ALL THE MATHEMATICAL SCIENCES ARE FOUNDED ON THE RELATIONS BETWEEN PHYSICAL LAWS AND LAWS OF NUMBERS.'

- James Clerk Maxwell

SCOTLAND
1831–1879

IDEAS AND INVENTIONS
Electromagnetic Radiation, Maxwell's Equations, Maxwell-Boltzmann Distribution, Colour Photography

FIELDS
Physics, Mathematics

Opposite: James Clerk Maxwell.

The Scottish physicist James Clerk Maxwell's breakthroughs in electromagnetism came largely in the early 1860s while he was a professor at King's College, London. He examined Faraday's idea concerning the link between electricity and magnetism interpreted in terms of fields of force and began to search for an explanation for this relationship. Maxwell soon saw that it was simple: electricity and magnetism were just alternative expressions of the same phenomena, a point he proved by producing intersecting magnetic and electric waves from a straightforward oscillating electric current. Furthermore, Maxwell worked out that the speed of these waves would be similar to the speed of light (186,000 miles per second) and concluded, as Faraday had hinted, that normal visible light was indeed a form of electromagnetic radiation. He argued that infrared and ultraviolet light were the same and predicted the existence of other types of wave – outside of known ranges at that time – which would be similarly explainable. The discovery of radio waves in 1888 by Heinrich Rudolph Hertz (1857–94) would later confirm this.

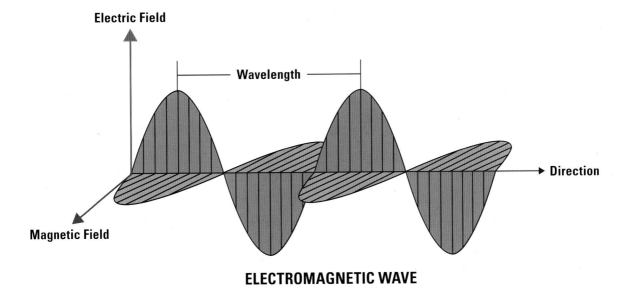

ELECTROMAGNETIC WAVE

James Clerk Maxwell greatly improved our understanding of electromagnetic waves.

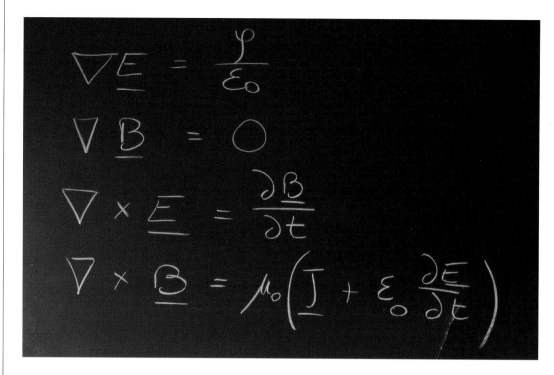

$$\nabla \underline{E} = \frac{\rho}{\varepsilon_0}$$

$$\nabla \underline{B} = 0$$

$$\nabla \times \underline{E} = \frac{\partial \underline{B}}{\partial t}$$

$$\nabla \times \underline{B} = \mu_0 \left(\underline{J} + \varepsilon_0 \frac{\partial \underline{E}}{\partial t} \right)$$

Maxwell's four equations.

But Maxwell did not stop there. In 1864, he published his 'Dynamical Theory of the Electric Field' which offered a unifying, mathematical explanation for electromagnetism. The text was based around four equations, now known simply as 'Maxwell's equations', which outlined the relationship between magnetic and electric fields. He later wrote another piece on this association, published in 1873 under the title *Treatise on Electricity and Magnetism*.

While Maxwell's most outstanding achievements were in explaining electromagnetic radiation, he also undertook important work in thermodynamics and would offer important kinetic explanations for the behaviour of gases. This involved building on the idea of the movement of molecules in a gas. The Scotsman proposed that the speed of these particles varied greatly. Again he used his mathematical skills to produce a statistical model which would reinforce the ideas behind this research, now known as the Maxwell-Boltzmann distribution

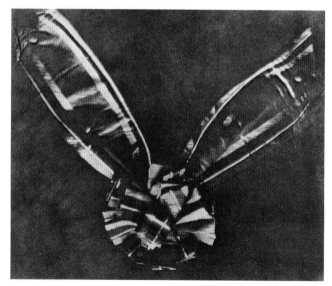

The first colour photograph, of a tartan ribbon, created by Maxwell in 1861.

law (the last part of the name coming from the Austrian Ludwig Eduard Boltzmann who independently discovered the same explanation). Amongst other things, the convincing explanation that heat in a gas is the movement of molecules would finally do away with the caloric fluid theory of heat.

Maxwell's other accomplishments involved the deduction that all other colours can be created from a mix of the three primaries. In 1861 he applied this discovery practically in photography, producing one of the first ever colour photographs.

Earlier in his career Maxwell had studied Saturn's rings and concluded that they were made up of lots of small bodies and could not be either a liquid or whole solid as had previously been speculated. In 1871, he returned to Cambridge and became the first Professor of Physics at the Cavendish Laboratory, which he helped to establish. The laboratory became world-renowned, dominating the progress of physics for many

Maxwell studied Saturn's rings and by creating a mathematical model for their motion, was able to show that they were made up of a series of small bodies rather than a whole liquid or solid.

decades and producing countless leading scientists.

It is highly possible that the Scotsman himself could have led many other breakthroughs had it not been for his tragically early death. He contracted and died from cancer, aged just 48.

Although regarded as a slow learner by his first tutors, William Hopkins, one of the country's brightest minds, recognised his great ability at university. Einstein also described the change in the conception of reality in physics that resulted from Maxwell's work as 'the most profound and the most fruitful that physics has experienced since the time of Newton'.

He may not be as well known, but James Clerk Maxwell's standing as a scientist is often considered by many to be on a par with Isaac Newton and Albert Einstein.

Like those other great scientists, he offered explanations for physical phenomena which would revolutionize our understanding of them. He forged a path for scientists to follow by taking the experimental discoveries of Faraday (1791–1867) in the field of electromagnetism and providing a unified mathematical explanation for an achievement which had evaded other minds for so long.

DMITRI
MENDELEEV

RUSSIA
1834–1907

IDEAS AND INVENTIONS
Periodic Table

FIELDS
Chemistry

Opposite: Dmitri Mendeleev.

'I WISH TO ESTABLISH SOME SORT OF SYSTEM NOT GUIDED BY CHANCE BUT BY SOME SORT OF DEFINITE AND EXACT PRINCIPLE.'

- Dmitri Mendeleev

The Pedagogical Institute in St Petersburg, where Mendeleev studied and then taught.

Mendeleev was born the last child of a large family in Tobolsk, Siberia, whose blind father could not support it. His mother enabled the family to survive by setting up and running a glass factory. Under such circumstances Mendeleev's early education was limited although he did attend school. He showed enough promise, though, to encourage his mother to leave Siberia in 1848, shortly after his father had died and the glass factory had burnt down, in a quest to enable him to enter university. Mendeleev was denied entry to a number of academic schools before settling to study science at the Pedagogic Institute in St Petersburg. Here he excelled and qualified as a teacher in 1855 before being given additional opportunities over the following years to study at universities in Russia and overseas.

In 1860 Mendeleev attended an important chemistry conference at Karlsruhe where the Italian, Stanislao Cannizzaro (1826–1910), passionately announced and backed his rediscovery of the distinction between molecules and atoms (originally made in 1811) by Avogadro (1776–1856). Understanding of the atomic weights of elements had been confused for half a century without this distinction and Cannizzaro's speech was a profound influence on the development of Mendeleev's later work.

During the 1860s Mendeleev returned to St Petersburg, finally becoming Chair of Chemistry at the university in 1866. He became aware during this period that chemistry lacked a comprehensive teaching textbook, so set about writing his own. This was finally published in 1869 under the title *The Principles of Chemistry*, setting a new standard. It was while he was researching this book that Mendeleev returned his attention to the atomic weights of elements, and introduced his card game into the equation.

Mendeleev wanted to list the known chemical elements in a structured way. Other scientists had tried to do the same in the past but were unsuccessful in finding a uniform way in which to list them or, indeed, in even deciding criteria by which to arrange them. So Mendeleev decided to write the properties of each element on a single card and began placing them in different formations according to various principles. He quickly discovered that if he positioned the elements according to atomic weight in short rows underneath each other, the resultant columns seemed to share common properties. The British chemist, John Alexander Reina Newlands (1837–98), had independently made a similar observation in 1864 but had had his observations ignored.

Mendeleev took his work a step further. He drew up a 'periodic table' of the elements according to their atomic weights and the common properties he found in the columns. He realized that for this scheme to work it was necessary to leave spaces for elements which he believed were as yet undiscovered. He could, though, predict their likely properties and weights

The Karlsruhe conference of 1860 established the distinction between atoms and molecules as scientific consensus. From this basis, Mendeleev was able to begin his work on the Periodic Table.

and was vindicated over the coming years when gallium, scandium and germanium were discovered to slot into the gaps he had left.

Mendeleev also believed that some atomic weights, such as that for gold, had been miscalculated and he re-estimated their details to fit his structure.

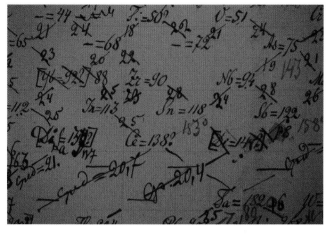

Mendeleev's hand-drawn periodic table.

Again, more accurate measurements would later prove Mendeleev's assumptions to be correct. He first published his table in 1869. The text was not widely accepted at first, but eventually it became the standard method of classifying the chemical elements, restructuring the entire subject of chemistry and greatly aiding scientists of all disciplines in understanding the properties and behaviour of the elements. Mendeleev predicted three yet-to-be-discovered elements including eke-silicon and eke-boron and his table did not include any of the Noble Gases which were still unknown. Mendeleev also investigated the thermal expansion of liquids, and studied the nature and origin of petroleum. In 1890 he resigned his professorship and in 1893 became Director of the Bureau of Weights and Measures in St Petersburg.

The element with the atomic number 101 was discovered in 1955 and named mendelevium as a tribute to the great Russian scientist who narrowly missed out on the Nobel Prize for Chemistry in 1906. He lost by one vote.

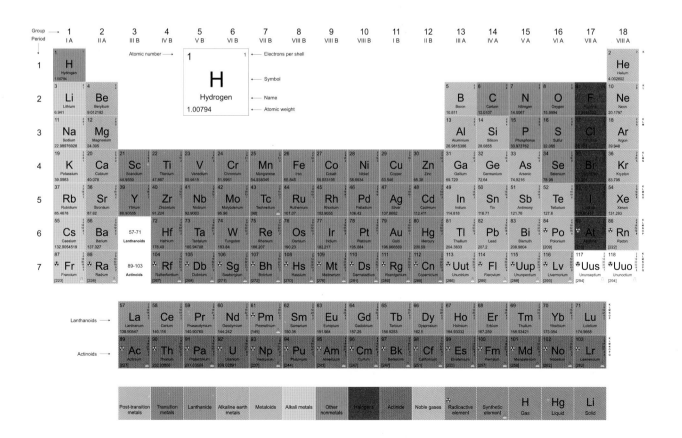

A modern version of the Periodic Table, including the elements predicted by Mendeleev, which have since been discovered.

WILHELM CONRAD RÖNTGEN

GERMANY
1845–1923

IDEAS AND INVENTIONS
Cathode Rays, X-Rays, Radiation

FIELDS
Physics, Chemistry

Opposite: Wilhelm Röntgen.

'HAVING DISCOVERED THE EXISTENCE OF A NEW KIND OF RAYS, I OF COURSE BEGAN TO INVESTIGATE WHAT THEY WOULD DO...'

- Wilhelm Röntgen

Wilhelm Röntgen in his laboratory, 1895.

Today the uses of X-rays, particularly in hospitals, are well known to the general public, yet little more than a century ago leading physicists were not even aware of their existence. It would take a chance discovery by a German named Wilhelm Conrad Röntgen to change that and begin a process which would not only result in an understanding of X-rays, but lead on to pioneering work in radioactivity.

Röntgen was a successful scientist in his own right long before he stumbled upon X-rays. He was a Professor of Physics at the University of Würzburg in Germany from 1888 and had conducted research in many areas. But he was largely unknown to the wider world until 28 December 1895 when he unfurled the exciting discovery for which he would subsequently be remembered. The story of the rays Röntgen named 'x', because of their mysterious properties on his first finding them, however, had actually begun a few weeks earlier in Röntgen's laboratory, on 8 November 1895.

Max von Laue.

He had been undertaking some tests involving little understood cathode rays when he noticed something unusual. He knew that the cathode rays emitted from the device he was using to project them could only travel a few centimetres, yet he suddenly noticed that another item in the darkened room became illuminated during the test. It was a screen covered in a substance called barium platinocyanide and Röntgen realized straight away that the glow could not have been caused by the cathode rays as the object was over a metre away. He thought that perhaps it indicated some unidentified radiation being emitted when the rays hit the glass wall of the projection device. He began excitedly investigating the properties of his accidental discovery.

Before announcing his finding to the world he uncovered many of the properties of the rays, including some of the factors which would go on to make them so useful in the future. For example, Röntgen discovered that the rays would pass through many different kinds of matter including metals, wood and, significantly, human limbs. Indeed, bones would appear as shadows against a screen or photographic plate allowing an X-ray image of them to be taken. He also found that the rays travelled in straight lines and were not knocked off course by electric or magnetic fields. But he was unsure exactly what the rays were; they had some characteristics in common with light rays, but did not reflect or refract like light. It was not until 1912 that they were fully understood, when Max Theodor Felix von Laue (1879–1960) showed that they were a form of electromagnetic radiation with a wavelength shorter than visible light.

The benefits of X-rays in medicine were quickly brought into common use, and as they were better understood they were applied to other areas, such as the study of the structure of molecules and in researching the properties of crystals. Other scientists ran into new phenomena as a by-product of their researches into X-rays, most notably Antoine-Henri Becquerel (1852–1908) who began to understand radioactivity as a result of his investigations. By the same token, it took time for the potentially harmful effects of X-ray radiation to be understood, and Röntgen's health was affected by his experiments.

Röntgen did, however, become an early beneficiary

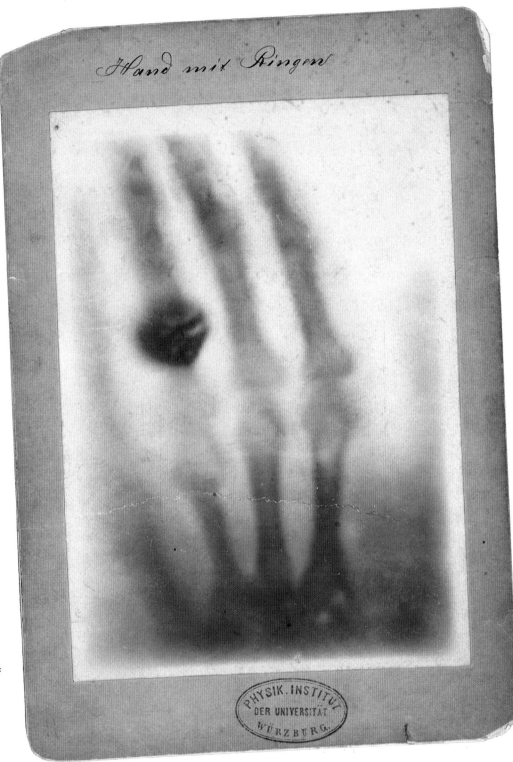

Hand mit Ringen

The first X-ray photograph taken by Röntgen was of his wife's hand.

of the Nobel Prize. In 1901 he was the first person to receive the award for physics in recognition of his discovery.

Wilhelm Röntgen was the first person to take X-ray photographs. His pictures included amongst other things images of his wife's hand.

After his discovery it only took him six weeks to determine many of the properties of X-rays.

The development was to be instrumental in the later discovery of radioactivity.

Röntgen also worked and researched in other scientific fields: elasticity, capillarity, the specific heat of gases, conduction of heat in crystals, piezoelectricity, absorption of heat by gases, and polarized light. Sadly, as a result of his experiments Röntgen and his technician were both affected by radiation poisoning.

PAUL
EHRLICH

*'WE MUST SEARCH FOR MAGIC BULLETS.
WE MUST STRIKE THE PARASITES, AND THE
PARASITES ONLY.'*

- Paul Ehrlich

GERMANY
1854–1915

IDEAS AND INVENTIONS
Bacterial staining, Cure for Syphilis, Chemotherapy, Antibodies

FIELDS
Biology, Medicine

Opposite: Paul Ehrlich.

After the work of Edward Jenner (1749–1823) and Louis Pasteur (1822–95), the role and value of vaccinations were widely realized in the fight against disease. By the start of the twentieth century, however, there remained many untreatable fatal illnesses. Scientists began looking for alternative ways of conquering disease. One who was particularly successful and who, in the process, founded a new approach to the discovery of cures, was the German Paul Ehrlich.

Earlier in his career, Ehrlich had been deeply impressed by the development of a new finding involving the 'staining' of cells to highlight them when studied under the microscope. Some of these dyes only stained particular types of microorganism and Ehrlich was instrumental in creating a dye which illuminated the tuberculosis bacillus discovered by Robert Koch (1843–1910) in 1882. This was an important achievement in itself, becoming a technique widely used in the diagnosis of tuberculosis.

A stained tuberculosis bacillus. Ehrlich was the first to create a dye that could stain tuberculosis bacteria and the technique has been used in diagnosis ever since.

The principles behind staining remained central to most of the other work undertaken by Ehrlich during the rest of his career, and would provide the inspiration for the achievement for which he is most remembered. From around 1905 Ehrlich began to thoroughly research his hypothesis that if a dye could latch solely onto harmful bacteria (as he had proved with his previous work on tuberculosis), then perhaps other chemicals would behave in a similar way. Instead of illuminating the disease-causing microorganisms, however, he hoped they would kill them. The chemical which would become the basis for the proof of Ehrlich's theory was arsenic. This was an element potentially fatal to humans, but which in certain compounds he found could be used effectively to kill bacteria without a harmful number of side effects. Ehrlich at last completed the successful trial of the 'magic bullet' as a treatment for disease in 1909. An arsenic-based compound that he had been testing hunted out and killed the organism which caused syphilis. The following year he launched his treatment under the name Salvarsan, and it was hugely popular in combatting the disease, a widespread and unpleasant affliction often resulting in insanity and death. Moreover, the technique Ehrlich had employed was regarded as the foundation of chemotherapy, the treatment of disease by the use of synthetic compounds to locate and destroy the organisms causing an illness. It was an approach which would go on to have vital importance in combatting so many other diseases, most noticeably cancer-causing cells.

In between his research on staining techniques and his cure for syphilis Ehrlich had jointly received a Nobel Prize for Physiology (in 1908) for a different discovery. From around 1889 to the turn of the century he was deeply involved with immunology and it was for this he received his award. He is often considered to be the founder of the modern approach to this area of science for his systematic and quantitative methods in attempting to understand it. He put forward theories on how the immune system worked and the role of antibodies. He also undertook a number of experiments designed to

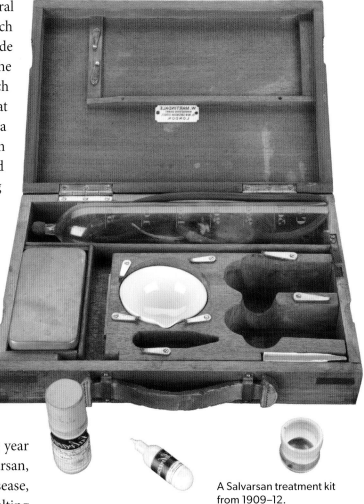

A Salvarsan treatment kit from 1909–12.

measure the increasing strength of the immune system in animals after repeated exposure to different types of disease-causing bacteria. This led to breakthroughs in the preparation of treatments for diphtheria and the development of techniques for measuring their effectiveness. Indeed, it was the later recognition of the limitation of these types of cures which would lead directly to Ehrlich's new approach to chemotherapy.

In our modern world with ready access to penicillin and other antibiotics, it is easy to forget the dreadful impact that diseases like smallpox and tuberculosis had on previous societies. Diseases which are now to all intents and purposes eradicated could spell a miserable death, even as recently as the 1950s. This is certainly the case with tuberculosis. In his obituary the London Times paid tribute to Ehrlich's achievement in opening new doors to the unknown, acknowledging that, 'The whole world is in his debt.'

Paul Ehrlich at work in his laboratory.

NIKOLA TESLA

CROATIA/USA
1856–1943

IDEAS AND INVENTIONS
Induction Motor, Alternating Current, Tesla Coil, Wireless Communication, Fluorescent Lights

FIELDS
Physics, Engineering

Opposite: Nikola Tesla.

'ERE LONG, INTELLIGENCE – TRANSMITTED WITHOUT WIRES – WILL THROB THROUGH THE EARTH LIKE A PULSE THROUGH A LIVING ORGANISM.'

- Nikola Tesla

Nikola Tesla, an eccentric electrical engineer, was born in modern day Croatia. The brilliant, eccentric and often troubled mind of Tesla was apparent from a young age. Although he did not come from an academic family background, there was a history of inventors in his ancestry and his father worked hard on developing Tesla's mental abilities. Despite interruptions to his childhood education due to frequent sickness and the severe trauma caused by the death of his older brother, Dane, Tesla progressed into higher education, taking up a place at the university of Graz in Austria.

While at the university, Tesla was exposed to demonstrations of existing generators and electric motors and began to ponder better ways of creating and transporting electricity. He later came up with an idea involving a rotating magnetic field in an induction motor which would generate an 'alternating current' (now known as AC). Most electricity being created at the time for use in homes, offices and factories involved a direct current (DC) which had its limitations, particularly the cost of generating it, its difficulty in being transported over long distances and its need for a commutator. By contrast, Tesla would later prove his alternating current could travel safely, efficiently and cheaply over long distances.

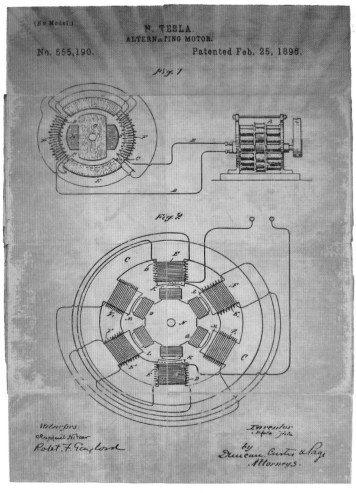

Nikola Tesla's patent on the alternating current motor from 1896.

His invention of an induction motor, in line with his earlier ideas in 1883 was the first big step on that road. His next move was to sell it.

Thomas Edison

Edison simply refused to accept that 'impossibilities' could not become facts without relentless experimentation and results which convinced him to the contrary. With some 1,093 patents singularly or jointly held in his name by the time he stopped inventing at 83, nobody could doubt that Edison meant what he said. He was the most prolific inventor the world had ever known, filing a patent once every two weeks of his working life. Some of his most significant inventions included the phonograph, the first ever sound-recording machine, designed and invented in 1877; the first commercially incandescent light bulb, produced in 1879 after more than 6,000 attempts at finding the right filament; and the Kinetograph and the Kinetoscope in 1894, which led to silent movies.

Telsa decided to emigrate to America, arriving penniless but soon finding work by making use of his electrical engineering skills. When he first landed in the USA in 1884, he was given work by the famous inventor Thomas Edison. Edison's revolutionary approach of establishing dedicated research and development centres full of inventors, engineers and scientists, working day and night on testing and building, brought many of his ideas to fruition. This began with his laboratory in Menlo Park, New Jersey, in 1876. Not only did these centres help Edison practically complete his own inventions but they also changed the rest of the business world's approach to research and development. But contrasting personalities and conflicting ideas between Edison and Tesla about electricity made the relationship a short one, sparking a bitter feud which would ultimately change the way the world received much of its power.

Within a year, Edison got rid of him. But Edison's rival, George Westinghouse, wooed Tesla. In 1885 his company, Westinghouse Electric, bought the rights to Tesla's alternating current inventions and a war of electricity began. Edison and others believed in and, probably more importantly, had a financial interest in, direct current and wanted to make it a success. It was already the standard way of generating and supplying electricity. Westinghouse and Tesla believed their method was ultimately more adapted to the job and fought hard to promote it. In spite of Edison's attempts at damaging the reputation of alternating current by claiming it was unsafe (which Tesla would later refute by grand demonstrations lighting lamps using only his body, by allowing his AC current to flow through it), the greater benefits of alternating current were soon realized. With the subsequent invention of better transformers in its transportation, alternating current became the standard, with DC increasingly confined to only specialist applications. The trend continues to this day.

George Westinghouse's display of Tesla's AC system at the 1893 Columbian Exposition.

A Tesla coil in operation.

In 1891, Tesla built on his knowledge to invent the 'Tesla coil' which was even more efficient at producing high frequency alternating current. It had many applications and still today is widely used in radio, television and electrical machinery. Using this and his turn of the century discovery of 'terrestrial stationary waves', which basically meant the planet earth could be employed as an electrical conductor, he produced some spectacular demonstrations. Tesla generated self-made 'lightning-strikes' over a hundred feet long, and he once lit 200 lamps, unconnected by wires, stretched over 25 miles. Indeed, the idea of the widespread transmission of electricity without wires became a particular area of interest for Tesla in his latter years. The 'tesla', the SI unit of magnetic flux density, is named in honour of him.

Tesla was also a prolific inventor. His inventions include: the telephone repeater, the rotating magnetic field principle, the polyphase alternating-current system, the induction motor, alternating-current power transmission, the Tesla coil transformer, wireless communication, radio, fluorescent lights, and more than 700 other patents.

Tesla's Wardenclyffe Tower project was an attempt at large-scale wireless power transmission. Financial difficulties led to the abandonment of the project in 1906 and the tower itself was demolished in 1917.

MAX
PLANCK

GERMANY
1858–1947

IDEAS AND INVENTIONS
Quantum Theory, Planck's Constant, Black Body Radiation

FIELDS
Physics

Opposite: Max Planck.

'MY RESEARCH ON THE ATOM HAS SHOWN ME THAT THERE IS NO SUCH THING AS MATTER ITSELF. WHAT WE PERCEIVE AS MATTER IS MERELY THE MANIFESTATION OF A FORCE.'

- Max Planck

When did the 'modern' scientific era actually begin? Throughout the nineteenth century, there had been many advances in all aspects of science which could arguably be seen to have launched a new foundation. But for physics at least, the answer is simple: 1 January 1900. That was the day the German Max Planck gave the first public enunciation, albeit to his son, of quantum theory. It was a notion which completely abandoned assumptions made in classical physics and it founded a whole new age.

To a degree, Planck stumbled across the concept that would change the scientific world through an element of chance. He had been undertaking theoretical work in thermodynamics and it was his search for a hypothetical answer to an inexplicable problem in physics at that time which led to a solution reflected in reality. The German, like many other scientists before him, had been considering formulae for the radiation released by a body at high temperature. He knew it should be expressible as a combination of wavelength frequency and temperature, but the 'irregular' behaviour of hot bodies made a consistent prediction difficult. For a theoretically perfect form of such a

matter known as a 'black body', therefore, physicists could not predict the radiation it would emit in a neat scientific formula. Earlier scientists had found expressions which were in line with the behaviour of hot bodies at high frequencies, and others found an entirely different equation to show their nature at low frequencies. But none could be found which fitted all frequencies and obeyed the laws of classical physics simultaneously.

Not that such a conundrum bothered Max Planck. Instead, he resolved to find a theoretical formula which would work mathematically, even if it did not reflect known physical laws. The answer he soon found was a relatively simple one: the energy emitted could be expressed as a straightforward multiplication of frequency by a constant which became known as 'Planck's constant' (6.6256×10^{-34} Js). But this only worked with whole number multiples (e.g. 1, 2, 3, etc.) which meant that for the formula to have any practical use at all, one had to accept the radical assumption that energy was only released in distinct non-divisible 'chunks', known as quanta, or for a single chunk of energy, a quantum. Up until

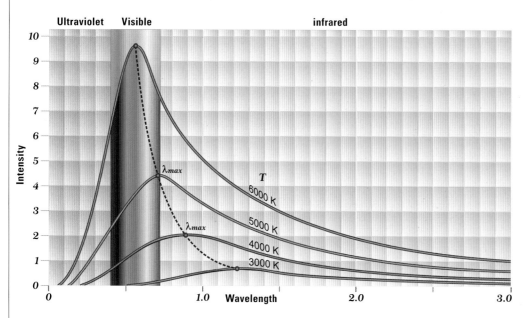

Max Planck established a theory of black body radiation showing the relationship between wavelength and temperature.

that point it had been assumed that energy was emitted in a continuous stream, so the idea it could only be released in quanta seemed ridiculous. It completely contradicted classical physics. But Planck's explanation fitted the behaviour of radiation being released from hot bodies. Moreover, the individual quanta of energy were so small that when emitted at the everyday, large levels observed in nature, it seemed logical energy could appear to be flowing in a continuous stream.

In this way classical physics was cast into doubt and quantum theory was born. Planck announced his results to the wider public in his 1900 paper 'On the Theory of the Law of Energy Distribution in the Continuous Spectrum'. It naturally caused a stir, but when Albert Einstein was able to explain the 'photoelectric' effect in 1905 by applying Planck's theory, and likewise Niels Bohr in his explanation of atomic structure in 1913, the notion suddenly

did not seem so ridiculous. The abstract idea really could explain the behaviour of physical phenomena and consequently Planck was quickly elevated to the status of Germany's most prominent scientist. He was awarded the Nobel Prize for Physics for his breakthrough in 1918.

The classical physics of Newton and Galileo provides us with laws capable of explaining the ordinary, everyday world around us. However, experiments conducted early in the twentieth century began to produce results that could not be explained by classical physics. One example was the discovery that if the electrons of an atom orbited the nucleus in the way classical physics predicted, they would spiral down into the nucleus within a very short time, and the atom would cease to exist. As this clearly was not the case, it became clear that another way of dealing with atomic and subatomic particles was required. This discovery, allied to Planck's quanta theory of energy, led to the development of quantum mechanics.

Opposite: Max Planck in his study.

GEORGE WASHINGTON
CARVER

USA
1864–1943

IDEAS AND INVENTIONS
Chemurgy, Permaculture, Synthetics

FIELDS
Agricultural Science

Opposite: George Washington Carver.

'I WENT TO THE TRASH PILE OF TUSKEGEE INSTITUTE AND STARTED MY LABORATORY WITH BOTTLES, OLD FRUIT JARS, AND ANY OTHER THING I FOUND THAT I COULD USE.'

- George Washington Carver

The American agricultural chemist George Washington Carver was arguably as much a teacher as a scientist. By disseminating his research on farming techniques and crops that would improve yields while conserving soil, and new uses for agricultural raw materials, he played a major role in diversifying and improving the agricultural economy of the South.

Carver was born into slavery in Missouri around 1864; the actual date is unknown. After a long struggle to access higher education, including rejection from one college due to his race, in 1894 he became the first African American to earn a Bachelor of Science degree in agricultural science at Iowa Agricultural College (now Iowa State University). His life experiences and Christian faith shaped his ambition to apply scientific research to the practical economic benefit of poor farmers in the South, many of whom were Black sharecroppers (tenant farmers) who couldn't afford modern machinery and fertilizers.

Having completed a Masters in Agriculture, in 1896 Carver became director of agricultural research at Tuskegee Institute (now Tuskegee University), the all-Black college in Alabama where he would spend his entire career. At its experimental farm he put his study of soil chemistry to work, testing methods to address the widespread soil depletion, low yields, and economic decline caused by decades of cultivating cotton as a single crop.

Carver demonstrated it was possible to significantly increase cotton yields over time by rotating cotton (or tobacco, another staple crop) with alternatives – in particular peanuts, soya beans, and sweet potatoes, which restored nitrogen levels to soil – and via techniques such as natural fertilization. As well as restoring nutrients, these protein-rich plants would improve farm workers' diets. In 1906 Tuskegee established a travelling agricultural school and laboratory to spread his message.

Carver was also aware of the need to grow commercial demand for surpluses of the alternative crops that he advocated. Early on in his tenure at Tuskegee, he founded a laboratory where he researched innovative food, industrial, and commercial applications. For example, sweet potato-derived products included stains, paints, ink, and rubber compounds, in addition to flours, starches, vinegar, and molasses. Carver regularly published his findings in bulletins such as 'How to build up worn-out soils' and 'Fertilizer experiments in cotton'. These bulletins often provided illustrated, practical guidance to farmers. In 1914, after a weevil infestation wrecked the cotton harvest, many farmers across America adopted his methods.

Carver came to nationwide fame in 1921 after persuading Congress to support US peanut growers by imposing a tariff on imports. His inspirational

George Washington Carver (second from right) teaching students at the Tuskegee Institute.

testimony on the huge number of potential uses for the legume earned him the nickname 'Peanut Man'. The American Chemical Society credited Carver with discovering more than 300 applications for peanuts. He developed peanut-derived products including milk, cooking oil, flour, cosmetic powder and lotions, massage oil, dyes, and wood stains, and experimented with medicines including antiseptics and

Carver earned the nickname 'Peanut Man' for the huge number of peanut-derived products he invented.

laxatives. He also listed nearly 120 applications for sweet potatoes, plus others for soybeans, pecans, and black-eyed peas.

As Carver did not rigorously record formulas from laboratory research, it is possible he collected and researched existing ideas as well as inventing his own original applications. He applied for only three patents for his plant products, including a peanut-based hair pomade and massage oil. The peanut was not a recognized crop when Carver began his experiments at Tuskegee; 50 years later it was the second most valuable cash crop after cotton in the South, and one of America's most important agricultural products. Carver's work helped to advance the understanding of the importance of soil conservation, agricultural training, and the benefits of a scientific approach to farming.

The Father of Chemurgy

Carver was inducted into the US Department of Agriculture's Hall of Heroes in 2000 as a pioneer of chemurgy, the process of extracting chemicals from agricultural raw materials to make industrial products. One early proponent, Henry Ford, established a chemurgy research laboratory in 1929 that developed soybean-based paints, plastic for car parts such as gearstick knobs, and even a concept 'soybean car' in 1940. As demand for rubber spiked and imports plummeted at the start of World War II, corn became an important ingredient in the production of synthetic rubber. Although attention shifted to oil and petroleum-derived industrial products by the 1950s, chemurgy could be making a comeback – especially as the fluctuating price of oil, its diminishing availability, and the environmental impact of its production make it a less than ideal source material to work with.

The soybean car, created by Henry Ford in 1940, used a plastic body made from soybeans and other agricultural products.

MARIE CURIE

POLAND
1867–1934

IDEAS AND INVENTIONS
Radioactivity, Discovery of Polonium and Radium

FIELDS
Physics, Chemistry

Opposite: Marie Curie.

'NOTHING IN LIFE IS TO BE FEARED; IT IS ONLY TO BE UNDERSTOOD.'

- Marie Curie

Quite aside from her practical achievements, Marie Curie is also important in the history of science for the pioneering role she played in opening up the subject to other women. She was arguably the first globally renowned and accepted female scientist and as such forged a path for all those of her sex who followed her. Her scientific discoveries in themselves were vital to understanding the new phenomenon of radioactivity. This was reflected in the fact that she was awarded not one, but two Nobel Prizes.

The majority of Curie's scientific work would take place in France, where she spent most of her life from

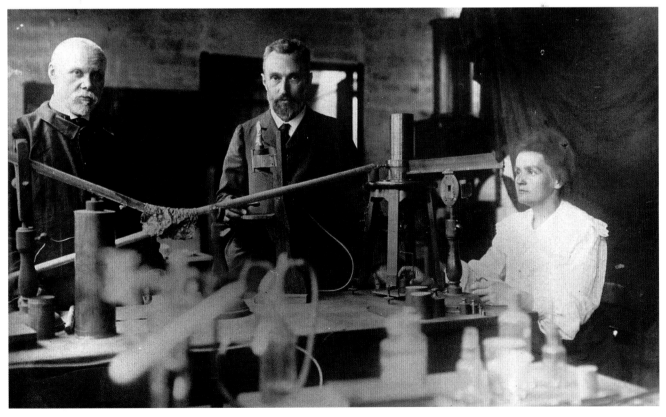

Marie and Pierre Curie in their laboratory with Henri Becquerel.

Pitchblende, the uranium ore from which Marie Curie was able to discover the elements of polonium and radium.

1891. Her country of birth, however, was Poland where she was born under the name Marya Sklodowska. Despite both her parents' status as teachers she grew up in relative poverty there. This was further accentuated when she was forced to move to Paris in order to obtain a higher education in physics, a level of study women were unable to undertake in her home country at the time. She graduated and shortly afterwards met her future husband, Pierre Curie (1859–1906), at the Sorbonne, where she studied and he worked. He was a respected physicist in his own right and it is no surprise that the two began working together in 1895, not long after they married.

The stimulus for the couple's later achievements would come initially from Marie's hunt for an area of research to undertake for her postgraduate studies. Encouraged by Pierre, she decided to further investigate the exciting, new discovery of radioactivity made by Henri Becquerel (1852–1908) in 1896. Curie's investigations into better understanding the properties of the phenomenon soon yielded results. Becquerel had proved that uranium was radioactive. Curie, wanting to find out which other elements were, quickly discovered that thorium shared similar traits. She went on to conclusively prove that radioactivity was an intrinsic atomic property of the element in question – uranium for example – and not a condition caused by other outside factors.

Curie's next achievement was to actually discover two new elements in 1898 through her researches, which she called polonium and radium, both highly radioactive, especially the latter. She had tracked down these elements after realizing uranium ore had a greater level of radiation than pure uranium, thereby correctly deducing that the ore must have

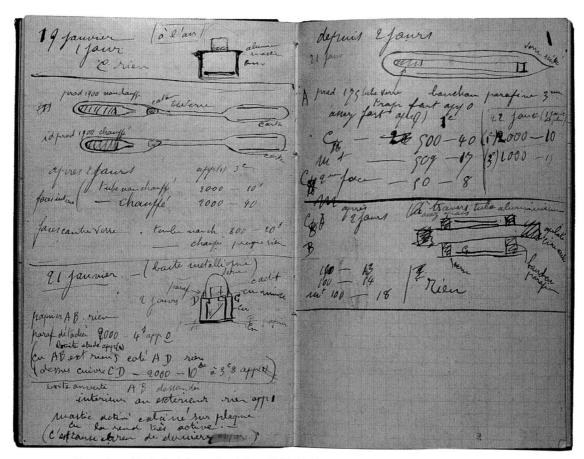

Pages from Marie Curie's notebook from 1899–1902 detailing her experiments on radiation.

Marie Curie's 1911 Nobel Prize.

contained other more radioactive, hidden elements. After these discoveries, Curie sought to obtain large enough quantities of the new substances to further understand their properties. Unfortunately, because of the minute amounts in which radium in particular was present in uranium ore, this meant she, along with her husband, had to wade through tonnes of the stuff for several years just to obtain a tenth of a gram by 1902. This, at least, allowed the calculation of the atomic weight of the new element to be made, as well as other work on its properties.

There was one question the Curies never fully got to the bottom of, however. What exactly was the radiation which came from these elements? Ernest Rutherford (1871–1937) would take the credit for the answer to this question with his explanation of 'alpha', 'beta', and later 'gamma', rays, but Marie did observe that the radiation was made up of at least two types of rays with distinct individual properties.

Sadly, Marie Curie would eventually die from leukaemia, which is thought to have been caused by her long exposure to radiation. At the time of her work with radioactive elements, the risks associated with radiation were not known and so no precautions were taken. Even to this day, her notebooks from her period of radioactive study remain too dangerous to examine.

Marie Curie would go on to be elected to her husband's former post of Professor of Physics at the Sorbonne, and become the institution's first female professor in the process. Her achievements in this position included the establishment of a research laboratory for radioactivity in 1912. The laboratory would go on to become world-renowned for its contribution to physics. This was due in no small part to a gift bestowed on Curie by the United States in 1921 which greatly facilitated the centre's work: a gram of the rare radium.

She received her second Nobel Prize in 1911 (her first was awarded with Pierre jointly in 1903), this time in chemistry, in recognition of her discovery of polonium and radium.

ERNEST
RUTHERFORD

'ALL SCIENCE IS EITHER PHYSICS
OR STAMP COLLECTING.'

- Ernest Rutherford

NEW ZEALAND
1871–1937

**IDEAS AND
INVENTIONS**
Radioactivity, Gamma Rays,
Atomic Nucleus, Planetary
Model of the Atom, Rutherford
Scattering, Proton Accelerators

FIELDS
Physics

Opposite: Ernest Rutherford.

After the discovery by Antoine-Henri Becquerel (1852–1908) of radioactivity in 1896, a number of scientists became responsible for a deeper understanding of the new phenomenon, including of course Marie Curie (1867–1934). However, the person who perhaps did most to bring a full understanding of radioactivity to the world, and greatly develop nuclear physics in general, was Ernest Rutherford.

The New Zealand-born scientist won a scholarship to the Cavendish Laboratory in Cambridge in 1895, working under the eminent J. J. Thomson (1856–1940). He went on to become a professor at the McGill University in Montreal in 1898. Here he put forward his observation that radioactive elements gave off at least two types of ray with distinct properties. These he named 'alpha' and 'beta' rays.

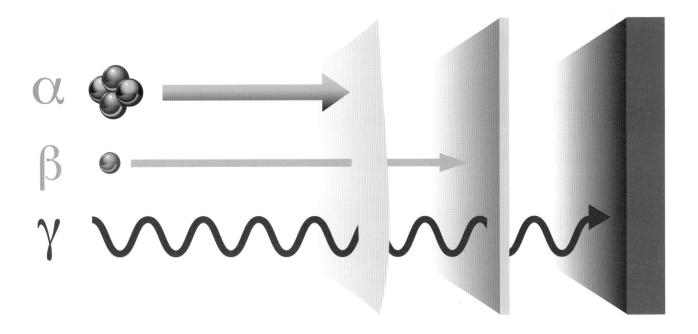

Ernest Rutherford discovered that radiation could be divided into alpha, beta and gamma rays.

In 1900 he confirmed the existence of a third type of ray, the 'gamma' ray, which was distinct in that it remained unaffected by a magnetic force, while alpha and beta rays were both deflected in different directions by such an influence.

It was also in Montreal that Rutherford met the British chemist Frederick Soddy (1877–1956). Between 1901 and 1903, the two collaborated on a series of experiments related to radioactivity and came to some startling conclusions. They showed how, over a period of time, half of the atoms of a radioactive substance could disintegrate through 'emanation' of a radioactive gas, leaving 'half-life' matter behind. The notion that atoms could simply decay away from an element was quite remarkable. Moreover, during the process the substance spontaneously transmuted into other elements – a revolutionary finding.

After the collaboration with Soddy ended, Rutherford went on to examine alpha rays more closely. He soon proved through experimental results that they were simply helium atoms missing two electrons (beta rays were later shown to be made up of electrons and gamma rays, actually short X-rays). During this period he moved back to England, to take up a post at Manchester University. Here he worked with Hans Wilhelm Geiger (1882–1945) to develop the Geiger counter in 1908. This device measured radiation and was used in Rutherford's work on identifying the make-up of alpha rays. He went on to use it even more significantly in his next major advance.

In 1910 Rutherford had proposed that Geiger and another assistant should undertake work to examine the results of directing a stream of alpha particles at a piece of gold foil. While most passed through and were only slightly deflected, about one in eight thousand bounced back virtually from where they had come! Rutherford was astonished, describing it later as 'quite the most incredible event that has ever happened to me in my life. It was almost... as if you had fired a fifteen-inch shell at a piece of tissue paper and it came back and hit you.'

He didn't let it fox him. In 1911 he put forward the correct conclusion: the reason for the rate of deflection was because atoms contained a minute nucleus which bore most of the weight, while the rest of the atom was largely 'empty space' in which

A Geiger counter from 1932. The Geiger counter was developed by Hans Geiger with the assistance of Ernest Rutherford.

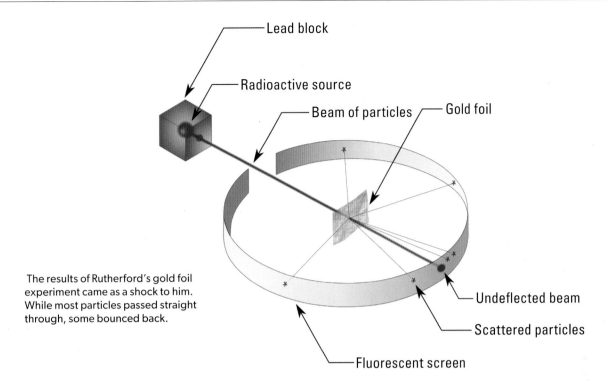

Lead block

Radioactive source

Beam of particles

Gold foil

Undeflected beam

Scattered particles

Fluorescent screen

The results of Rutherford's gold foil experiment came as a shock to him. While most particles passed straight through, some bounced back.

electrons orbited the nucleus much as planets did the Sun. The reason that the one in eight-thousand alpha particles bounced back was because they were striking the positively charged nucleus of an atom, whereas the rest simply passed through the spacious part. It was a vital discovery on the path to understanding the construction of the atom and would greatly aid Niels Bohr in his related revelations of 1913.

During World War I Rutherford served in the British Admiralty. Afterwards, in 1919, he was appointed to the chair of the Cavendish Laboratory at Cambridge. In the same year he made his final major discovery. Working in collaboration with other scientists, he found a method by which he could artificially disintegrate an atom by inducing a collision with an alpha particle. Essentially, what we now know as protons could be forced out of the nucleus by this smash. In the process the atomic make up of the substance changed, thereby transforming it from one element to another.

In this first instance he transmuted nitrogen into oxygen (and hydrogen), but went on to repeat the process with other elements.

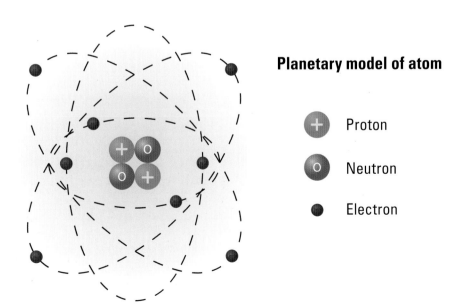

Planetary model of atom

+ Proton

O Neutron

● Electron

Rutherford's planetary model of the atom was hugely influential.

ALBERT EINSTEIN

'SCIENCE WITHOUT RELIGION IS LAME;
RELIGION WITHOUT SCIENCE IS BLIND.'

- Albert Einstein

GERMANY/USA
1879–1955

IDEAS AND INVENTIONS
Photoelectric Effect, General and Special Relativity, Spacetime, E=mc², Gravitational Waves, Cosmological Constant

FIELDS
Physics

Opposite: Albert Einstein.

Of the essays written by Einstein in 1905, arguably the most influential was his enunciation of a 'special' theory of relativity, which advanced the idea that the laws of physics are actually identical to different spectators, regardless of their position, as long as they are moving at a constant speed in relation to each other. Above all, the speed of light is constant. It is simply that the classical laws of mechanics appear to be obeyed in our normal lives because the speeds involved are insignificant.

But the implications of this principle if the observers are moving at very different speeds are bizarre and normal indicators of velocity such as distance and time become warped. Indeed, absolute space and time do not exist. Therefore if a person were theoretically to travel in a vehicle in space close to the speed of light, everything would look normal to them but another person standing on earth waiting for them to return would notice something very unusual. The space ship would appear to be getting shorter in the direction of travel. Moreover, whilst time would continue as 'normal' on earth, a watch telling the time in the ship would be going slower from the earth's perspective even though it would seem correct to the traveller

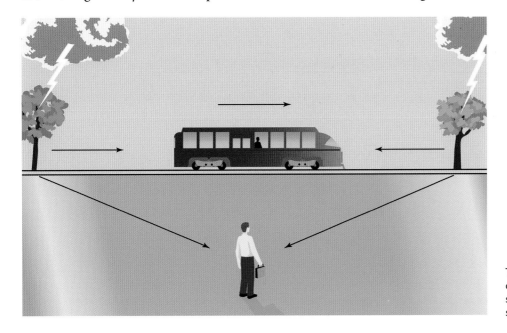

The lightning strike may appear to occur at different times for the man standing on the grass and the man sitting on the bus.

(because the faster an object is moving the slower time moves). This difference would only become apparent when the vessel returned to earth and clocks were compared. If the observer on earth were able to measure the mass of the ship as it moved, he would also notice it getting heavier too. Ultimately nothing could move faster than or equal to the speed of light because at that point it would have infinite mass, no length, and time would stand still!

From 1907 to 1915, Einstein developed his special theory into a 'general' theory of relativity which included equating accelerating forces and gravitational forces. Implications of this extension of his special theory suggested light rays would be bent by gravitational attraction and electromagnetic radiation wavelengths would be increased under gravity. Moreover mass, and the resultant gravity, warps space and time, which would otherwise be 'flat', into curved paths which other masses (eg, the moons of planets) caught within the field of the distortion follow.

Amazingly, Einstein's predictions for special and general relativity were gradually proven by experimental evidence. The most celebrated of these was the measurement taken during a solar eclipse in 1919 which proved the sun's gravitational field really did bend the light emitted from stars behind it on its way to earth. It was the verification which led to Einstein's world fame and wide acceptance of his new definition of physics.

Einstein spent much of the rest of his life trying to create a unified theory of electromagnetic, gravitational and nuclear fields but failed. It was at

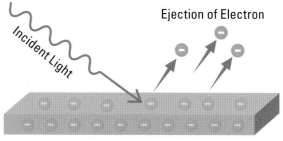

PHOTOELECTRIC EFFECT

Ejection of Electron

Incident Light

Metal Plate

Albert Einstein won the 1921 Nobel Prize for his discovery of the photoelectric effect. When light hits a metal, some electrons are emitted. From this, Einstein was able to calculate the energy of a light quantum.

least in keeping with his own remark of 1921 that 'discovery in the grand manner is for young people and hence for me is a thing of the past'.

Fortunately, then, he had completed three other papers in his youth (in 1905) in addition to his one on the special theory of relativity! One of these included the now famous deduction which equated energy to mass in the formula $E=mc^2$ (where E=energy, m=mass and c=the speed of light). This understanding was vital in the development of nuclear energy and weapons, where only a small amount of atomic mass (when released to multiply by a factor of the speed of light squared under appropriate conditions) could unleash huge amounts of energy. The third paper described Brownian motion, and the final paper made use of Planck's quantum theory in explaining the phenomenon of the 'photoelectric' effect, helping to confirm quantum theory in the process.

Almost inevitably, Einstein was also drawn into the atomic bomb race. He was asked by fellow scientists in 1939 to warn the US President of the danger of Germany creating an atomic bomb. Einstein himself had been a German citizen, but had renounced his citizenship in favour of Switzerland, and ultimately America, having moved there in 1933 following the elevation of Hitler to power in his home country. Roosevelt's response to Einstein's warning was to initiate the Manhattan project to create an American bomb first.

After the war, Einstein spent time trying to encourage nuclear disarmament.

Einstein argued that space and time were part of the same phenomenon as part of his theory of relativity. Gravity would distort the otherwise 'flat' spacetime.

Einstein is best known for his discovery of the theory of relativity, but it was actually his work on the photoelectric effect that won him the Nobel Prize.

ALEXANDER FLEMING

SCOTLAND
1881–1955

IDEAS AND INVENTIONS
Penicillin

FIELDS
Biology, Medicine

Opposite: Alexander Fleming.

'ONE SOMETIMES FINDS WHAT ONE IS NOT LOOKING FOR.'

- Alexander Fleming

Alexander Fleming had had a quite unremarkable life up until the chance discovery of a mould in his laboratory in September 1928. Even the decade after he made the find which would go on to save millions of lives, little changed. It took until 1940 for the 'penicillin' to be produced from the fungi in practically useful quantities. Although it was in fact another team of people altogether who facilitated the latter development, it was Fleming who became revered as a hero.

The son of a Scottish farmer, Fleming came from a humble background, and began his working life at the age of sixteen as a shipping clerk in London, England. After inheriting a small amount of money, and following suitable encouragement from his brother who was a doctor, Fleming decided to study medicine. In 1902, he joined St Mary's Hospital Medical School in London, where he remained for the rest of his career barring a period from 1914–18 putting his medical skills to good use for the war effort.

Fleming became increasingly interested in bacteriology. Indeed, it was his wartime experiences which made him realize there was a need for a non-toxic drug to combat the millions killed by the bacteria which infected wounds. After he rejoined St Mary's therefore, he searched for a naturally occurring bacteria-killer and focused initially on what he believed were the body's own sources: tears, saliva and

Fleming working with penicillin in his laboratory.

mucus from the nose. In 1922, he had his first success, producing lysozyme, an enzyme produced by the body. It killed certain bacteria naturally, but Fleming could not produce it in sufficiently concentrated quantities to be of medical use.

The search continued, although even scientists sometimes have to take a holiday, and ironically it was a two-week break which led to Fleming's ultimately world-changing discovery. Before leaving for his vacation in 1928, however, the bacteriologist had been examining some dishes containing staphylococcus bacteria, which turned out to be the first in a sequence of rather fortunate events. He accidentally left one of the dishes exposed to the air before he departed and it became infected with Penicillium notatum. The form of infection in itself was lucky as it was only because it was being studied elsewhere in the hospital that it was present to contaminate Fleming's sample at all, and a cold spell of weather in Fleming's absence allowed the fungi which developed to grow.

Although Fleming had luck on his side in the first instance, the fact he was a skilled bacteriologist was also vital. On returning from his holiday, he noticed a mould had grown in the infected dish and, rather than simply wash it out, was sufficiently interested to examine it further. He noticed clear patches around the edges of the contamination and correctly deduced that there was something in the Penicillium notatum which was killing the staphylococcus bacteria. On further

testing he found it was a useful killer of many forms of bacteria, but again it occurred in insufficient quantities to be of much further use.

Natural and semi-synthetic versions of penicillin would go on to be mass-produced, saving millions of lives during the war and even more afterwards as it was used to combat a whole series of bacteria-causing diseases. Fleming would be hailed as a saviour by a public in need of heroes and was knighted in 1944, although the team of later scientists had arguably done most to make penicillin useful. Fleming himself said of his role, 'My only merit is that I did not neglect the observation and that I pursued the subject as a bacteriologist.'

So it was left to the spur brought about by World War II over a decade later and a new team of scientists before the quest for a non-toxic antibiotic was revived and 'penicillin' as Fleming had named his finding was revisited. The Scot supplied the team led by Howard Walter Florey and which included a chemist called Ernst Boris Chain with a sample of his mould. By 1940 the team had proved penicillin's potency in fighting infections in mammals and soon afterwards made the breakthroughs necessary for it to be produced on an industrial scale.

The importance of Florey and Chain to the story was at least acknowledged when, along with Fleming, they were jointly awarded the Nobel Prize for Physiology in 1945.

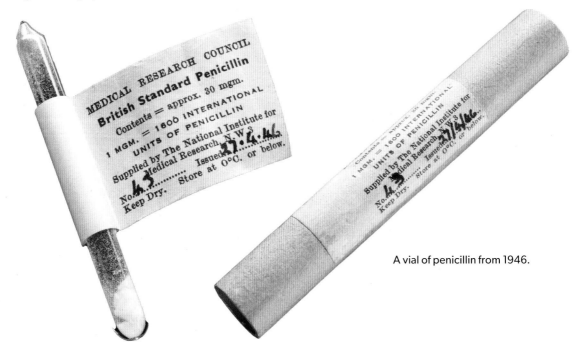

A vial of penicillin from 1946.

A penicillin production line during World War II. The war spurred the development of mass production of penicillin. Before the war, it had languished as a scientific curiosity and its true importance had not been realized.

NIELS BOHR

DENMARK
1885–1962

IDEAS AND INVENTIONS
Bohr Model of the Atom,
Correspondence Principle,
Complementarity Principle,
Nuclear Fission

FIELDS
Physics

Opposite: Niels Bohr.

'THOSE WHO ARE NOT SHOCKED WHEN THEY FIRST COME ACROSS QUANTUM THEORY CANNOT POSSIBLY HAVE UNDERSTOOD IT.'

- Niels Bohr

Few twentieth century theoretical physicists are referred to in the same breath as Albert Einstein (1879–1955) but the Dane Niels Bohr is one of them. He made major contributions to validating the concept of quantum physics set out by Max Planck (1858–1947) in 1900, solved issues concerning the behaviour of electrons in Ernest Rutherford's atomic structure and was involved in the development of the first atomic bomb.

Bohr gained his PhD at the University of Copenhagen in 1911 then moved briefly to the Cavendish Laboratory at Cambridge before settling in Manchester to work with Ernest Rutherford. The New Zealand physicist had just established his 'planetary' model of the atom: a tiny central nucleus bore most of the weight, around which electrons spiralled in a series of orbits. But there was a problem with this model. Classical physics insisted that if the electrons moved around the atom in this way, the energy they radiated would ultimately expire and the electrons would collapse into the nucleus. In 1913 Bohr resolved this issue and simultaneously validated Rutherford's model by applying Planck's quantum theory to it. He argued, from the perspective of quantum theory, that electrons only existed in 'fixed' orbits where they did not radiate energy. Quanta of radiation would only ever be emitted as an atom made the transition between states and absorbed or released energy. Only

at this point in time would electrons 'move', hopping from a lower to a higher-energy orbit as the atom took on energy, or jumping down an orbit as it emitted it (producing light in the process).

Bohr calculated the amount of radiation emitted during these transitions using Planck's constant. It fitted physical observations. Also, when he applied this to hydrogen atoms and the wavelengths of light that they should have released under this principle, he again found his calculations matched real world examples. It was a bizarre concept to grasp, as had been Planck's initial enunciation of quantum theory, but here was another practical example which validated it.

Bohr made another important addition to quantum theory and one to the school of quantum mechanics which succeeded it. In the former in 1916 he enunciated the 'correspondence principle': in spite of the huge apparent differences between the two, the laws which govern quantum theory at the microscopic level should still correspond with our understanding of classical physics as observed on the larger 'real world' level.

Later in 1927, in quantum mechanics, Bohr added the 'complementarity principle'. This argued that debates over whether light, as well as other atomic objects, behaved in a wave-like or particle-like fashion were futile because the equipment used in experiments to try to prove the case one way or the

An electron is accompanied by an emitted or absorbed amount of electromagnetic energy through its jumps between energy levels (a.k.a. "orbits").

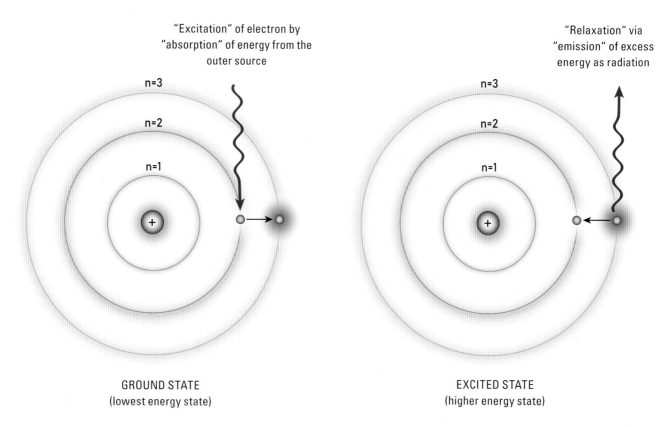

"Excitation" of electron by "absorption" of energy from the outer source

n=3
n=2
n=1

GROUND STATE
(lowest energy state)

"Relaxation" via "emission" of excess energy as radiation

n=3
n=2
n=1

EXCITED STATE
(higher energy state)

When the electron gets moved from its original energy level to a higher one, it then jumps back each level until it comes to the original postion, which results in a photon being emitted.

Nuclear Fission or Splitting

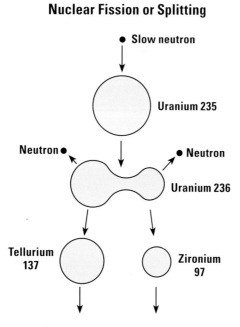

● Slow neutron

Uranium 235

Neutron ●   ● Neutron

Uranium 236

Tellurium 137   Zironium 97

Radioactive decay to other products

Above: The Bohr model of the atom suggested that electrons emitted quanta of radiation when they moved between different energy levels.

Left: Niel Bohr's liquid-drop model. The entry of a slow neutron into the nucleus of uranium-235 would cause it to split like a drop of water into two smaller droplets, emitting neutrons while doing so – forming the basis of nuclear fission.

Opposite: Niels Bohr in his laboratory.

other greatly influenced the outcome of the results. Instead all results only gave a partial glimpse of the answer to any atomic test and therefore had to be interpreted side-by-side with all other results to give a broader 'sum-of-the-parts' understanding. This idea sat neatly beside theories offered by the likes of Louis de Broglie (1892–1987), Werner Heisenberg (1901–76) and Max Born (1882–1970).

Bohr's theoretical and practical involvement in the physics which led to the creation of the first atomic bomb was dramatic. In 1939 he developed a theory of nuclear fission (splitting a heavy atom's nucleus to release huge amounts of energy suitable for an atomic bomb) with John Archibald Wheeler (1911–2008) from Bohr's 'liquid-drop' description (1936) of the way protons and neutrons bonded in the nucleus. He realized, ominously, that the uranium-235 isotope would be more susceptible to fission than the more commonly used uranium-238. Ultimately his findings would make their way to the US atomic bomb project, especially after Bohr escaped to America to flee occupied Denmark and acted as a consultant to the team.

Bohr was, however, uncomfortable with the implications of the new technology and dedicated much of the rest of his life to encouraging the control and limitation of nuclear weapons, founding the Atoms for Peace Movement for physicists with similar opinions to his own. Bohr was awarded the Nobel Prize for Physics in 1922.

ERWIN SCHRÖDINGER

'QUANTUM PHYSICS THUS REVEALS A BASIC ONENESS OF THE UNIVERSE.'

- Erwin Schrödinger

AUSTRIA
1887–1961

IDEAS AND INVENTIONS
Quantum Wave Mechanics, Schrödinger's Cat, Quantum Entanglement, Superposition

FIELDS
Physics

Opposite: Erwin Schrödinger.

The mid-1920s was open season in the field of quantum theory and one of the many physicists who waded in with a new influential direction was Erwin Schrödinger. Born in Vienna, to a prosperous merchant family, Schrödinger had a grandmother who was half Austrian, and half English, the English side of the family originating from Leamington Spa, and he grew up speaking both English and German in the home. Schrödinger was taught at home by a private tutor until he was ten. The Austrian scientist developed what became known as 'wave mechanics', although like others, including Einstein, he later became uncomfortable with the direction quantum theory took after doing so much in the first place to validate it.

Schrödinger's own development was built largely on the back of the 1924 proposal by Louis de Broglie (1892–1987) that particles could, in quantum theory, behave like waves. Whilst the Austrian Schrödinger was attracted to this explanation, he was troubled by certain implications of it. Essentially, he felt de Broglie's equations were too simplistic and did not offer a detailed enough analysis of the behaviour of matter, particularly at the subatomic level. So he took things a stage further and removed the idea of the particle completely! In its place, he argued everything was a form of wave.

Amazingly, between 1925 and 1926 he was able to calculate a 'wave equation' which mathematically underpinned this argument and the science of quantum wave mechanics was born. Further proof came when the theory was applied against known values for the hydrogen atom, and correct answers were obtained, for example, in calculating the level of energy in an electron. It clearly overcame some of the more woolly elements of the earlier quantum theory developed by Niels Bohr (1885–1962) and addressed the weaknesses in de Broglie's thesis.

Indeed, the theory behind wave mechanics was now applied to all sorts of other situations with great effect. Unfortunately, it too had some fundamental weaknesses and Schrödinger was not blind to these. The overriding one was, having done away with particles, it was difficult to offer a physical explanation for the properties and nature of matter. The Austrian came up with the concept of 'wave packets' which would give the impression of the particle as seen in classical physics, but would actually be a wave. The justifications he offered, though, were found not to add up.

This left Schrödinger's work susceptible to being superseded by that of others, just as his had improved on those whose ideas came before him. Shortly afterwards, the probabilistic interpretation of quantum

153

Hydrogen Wave Function
Probability density plots.

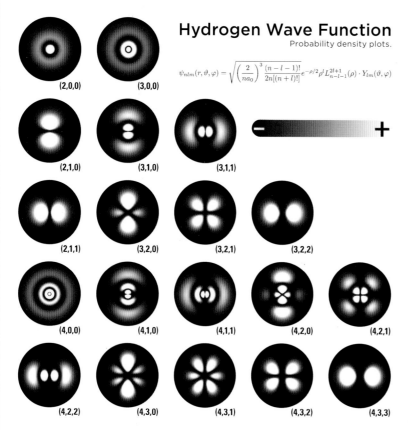

$$\psi_{nlm}(r,\vartheta,\varphi) = \sqrt{\left(\frac{2}{na_0}\right)^3 \frac{(n-l-1)!}{2n[(n+l)!]}} e^{-\rho/2} \rho^l L_{n-l-1}^{2l+1}(\rho) \cdot Y_{lm}(\vartheta,\varphi)$$

Hydrogen density plots, representing the probability of electron locations in hydrogen atoms. The brighter areas show the highest likelihood of finding an electron.

theory based on the ideas of Heisenberg (1901–76) and Born (1882–1970) took hold. This effectively proposed matter did not exist in any particular place at all, being everywhere at the same time, until one attempted to measure it. At that point the equations they put forward offered the best 'probability' of finding the matter in a given location. Whilst this is still widely accepted as the most adequate explanation today, Schrödinger joined Einstein and others in condemning such a loose, probabilistic view of physics where nothing was explainable for certain and essentially cause and effect did not exist.

Ironically, Paul Adrien Maurice Dirac (1902–84), another important influence in quantum mechanics, went on to prove Schrödinger's wave thesis and the alternative probabilistic interpretation he abhorred were mathematically, at least, the equivalent of each other. Schrödinger shared a Nobel Prize for Physics in 1933 with Dirac.

Schrödinger's Cat

This famous 'animal' is in fact part of a thought experiment designed by Erwin Schrödinger in the 1930s, to try and explain the problem that, contrary to all logic, atoms can exist in two states simultaneously, decayed and undecayed. Schrödinger uses the analogy of a cat locked in a box with a vial containing deadly poison. The vial's lid contains a radioactive atom. If the atom decays, the released particle opens the vial and the cat dies. This is an example of a quantum system, in which it seems the cat exists in an indeterminate state, because the atom is both decayed and undecayed, implying that the unobserved cat is neither dead nor alive, which is patently absurd.

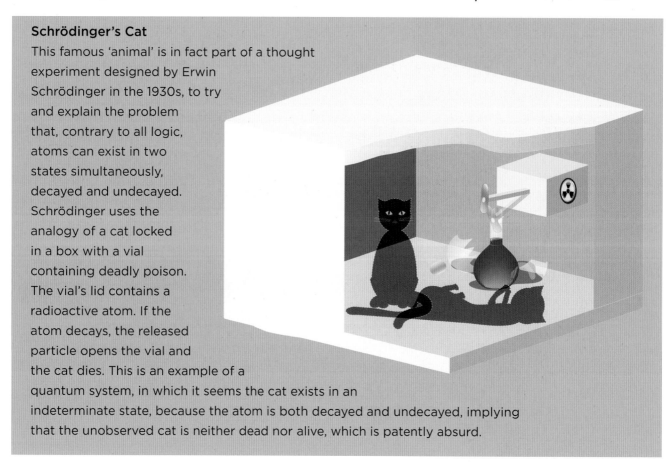

Louis de Broglie provided the foundations for the construction of Schrödinger's theory of quantum waves.

EDWIN HUBBLE

USA
1889–1953

IDEAS AND INVENTIONS
Galaxy Classification, Hubble's Constant, Hubble's Law, Expanding Universe

FIELDS
Astronomy

Opposite: Edwin Hubble.

'I KNEW THAT EVEN IF I WERE SECOND OR THIRD RATE, IT WAS ASTRONOMY THAT MATTERED.'

- Edwin Hubble

The man who completely changed our view of the bubble in which we exist was almost lost to astronomy, first to boxing, then to law. The young Hubble was such a fine fighter during the days of his astronomy and mathematics degree at the University of Chicago boxing promoters tried to persuade him to turn professional. He refused the offer. He did not turn down the chance to go to Oxford University in the United Kingdom on a Rhodes Scholarship to study law in 1910, though. He duly gained a BA in 1912 and contemplated a career in law on returning to the United States. In comparison to astronomy, he found the subject boring, however, so instead returned to Chicago to gain his PhD in the field of study he loved. After serving and being injured in World War I, he finally had the chance to observe the stars professionally, taking up a post in the Mount Wilson Observatory in California in 1919, where he would spend the rest of his career.

The astronomer was lucky in that shortly after he arrived, the observatory built a brand new 100-inch telescope, which was the most powerful in the world at that time. It allowed Hubble to view the skies in a level of previously unseen detail. He quickly took full advantage of this privilege. The American was particularly interested in the many 'nebulae' in the skies, all of which were thought to be clouds of dust within our own Milky Way galaxy. Indeed, it was thought at the time there was only this one galaxy in all, which according to the measurements of Hubble's contemporary and rival, Harlow Shapley (1885–1972), was approximately 300,000 light years across (this was subsequently revised to 100,000 light years).

Focusing on the Andromeda nebula, Hubble used a technique developed by Shapley himself to ascertain that this 'cloud' was some 900,000 light years away from earth and therefore clearly outside the Milky Way. Moreover, Hubble soon came to realize these spiral-shaped nebulae were in fact other galaxies, much like our own. There were literally millions of them in the sky, containing billions of other stars. The results were breathtaking, completely changing our perception of the size of the universe, and brought Hubble fame overnight.

Moreover, during the next few years, Hubble continued measuring the distances of the galaxies from earth and found they seemed to be moving away from it, or 'receding'. In addition, the greater the

The 100-inch telescope at the Mount Wilson Observatory used by Hubble to locate galaxies beyond the Milky Way.

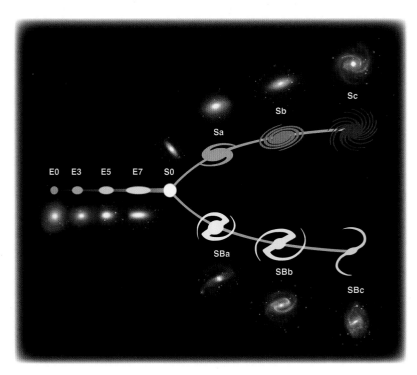

Hubble's galactic classification system is still used today.

this alteration, on hearing the universe was actually expanding, as 'the greatest blunder of my life'.

By 1929, Hubble had measured the distances of enough galaxies to announce his formulation of 'Hubble's constant'. He had worked out the speed at which the galaxies were recessing to be distance multiplied by his constant. Although Hubble overestimated the size of the constant, his formula was valid. Corrections since have allowed astronomers to estimate the radius of the universe to be a maximum of 18 billion light years and its age to be between 10 to 20 billion years old. Hubble went on to provide a system of classifying galaxies which is still largely in use.

distance between the earth and the galaxy, the faster the latter seemed to be receding. By 1927, Hubble came to the only logical conclusion: the universe, which most astronomers had believed was static, was in fact expanding. Other scientists had for the first time hinted at this possibility a few years earlier but now Hubble had provided conclusive evidence. Indeed, Einstein himself had developed an earlier theory which required the universe to be moving either inwards or outwards for it to work, but had changed it because astronomers had told him the universe was definitely static. He later referred to

Edwin Hubble, most widely known today because of the space telescope named after him, revolutionized our understanding of the cosmos. In the same way that that telescope hoped to improve our perception of the universe after its launch in 1990, so the American provided the most incredible 'picture' of space that humans had ever known, some sixty-five years earlier.

The notion of a universe which was expanding allowed later scientists to, amongst other things, find consensus on the origin of space and settle on the big bang theory. Indeed, the principle of an expanding cosmos has been at the heart of astronomical theory ever since.

The Hubble Space Telescope, named in honour of the great astronomer.

ENRICO FERMI

ITALY
1901–1954

IDEAS AND INVENTIONS
Radioactive Beta Decay,
Nuclear Fission, Fermi Paradox

FIELDS
Physics

Opposite: Enrico Fermi.

'IF THE RESULT CONFIRMS THE HYPOTHESIS, THEN YOU'VE MADE A MEASUREMENT. IF THE RESULT IS CONTRARY TO THE HYPOTHESIS, THEN YOU'VE MADE A DISCOVERY.'

- Enrico Fermi

Enrico Fermi, arguably Italy's most talented scientist of the twentieth century and quite possibly since Galileo, could have had no idea of the eventual outcome of the experimental work he undertook in Rome in the mid-1930s. He was systematically working his way through the elements to study the effects on them of a neutron-bombarding technique he had discovered. Most yielded predictable, or certainly not extraordinary, results. When he arrived at uranium, the heaviest naturally occurring element, however, something very odd happened, which was to have enormous impact on physics and beyond.

A few years later, in Chicago, Fermi would experience first hand the potential of his discovery. Fermi and his Jewish wife had fled to America following the rise of anti-Semitism in Italy.

Shortly afterwards, he had received reports of a reinterpretation of his uranium bombardment experiment. Fermi himself had been unsure of what had happened, suspecting the possibility that perhaps the uranium had transmuted into new, heavier elements. Now, however, an alternative explanation was offered by the German scientists Otto Hahn, Fritz Strassmann and Lise Meitner that the uranium nucleus had in fact been broken down into a number of smaller elements. Moreover, this nuclear fission had seen some of the uranium mass converted into potentially huge amounts of energy under the rules of Einstein's formula $E=mc^2$. This reinterpretation was leaked out of Germany by Meitner and her nephew Otto Frisch when they escaped the Nazi state.

Fermi immediately saw the impact of the analysis and set to work on reproducing the experiment with Niels Bohr on arrival in the US. They confirmed their best and worst fears: using the uranium isotope-235 a nuclear chain reaction could almost certainly be created as the basis of an atomic bomb. Fermi was recruited to the Manhattan Project to ensure the US created a fission bomb ahead of the Germans. Fermi led a team in Chicago seeking to generate a self-sustaining, contained nuclear reaction. By 2 December 1942 his team had created an 'atomic pile' of graphite blocks, drilled with uranium which went on to produce a self-sustaining chain reaction for nearly half an hour. 'The Italian navigator,' as one commentator reported back to the project committee, 'has just landed in the new world.' Less than three years later, the technology would be used in the first atomic bombs with devastating effect.

The Chicago pile, the world's first nuclear reactor.

by showering certain elements with alpha particles.

Fermi had quickly realized that the newly discovered neutrons would be even more suited to this purpose because their neutral charge would be more likely to allow them to slip into elements' nuclei without resistance. By chance he also found the phenomenon of 'slow neutrons' by placing a piece of solid paraffin in front of his target element during bombardment. This had the effect of slowing the neutrons down before they reached the element, meaning they were exposed to its nuclei for longer and thereby had a much greater chance of being drawn in to create new isotopes. As Fermi now worked through the elements applying these discoveries, he created lots of new radioactive isotopes, which was considered achievement enough for him to be awarded the 1938 Nobel Prize for physics. It was only after he had collected his award that the much more significant consequences of this work when applied to uranium were realized.

The innocent discovery back in Italy in the 1930s, which had led to such incredible consequences, had been Fermi's conception of neutron bombardment in the artificial transmutation of elements. The Juliot-Curies had announced in 1934 their discovery that radioactive isotopes could be generated artificially

Earlier in his career, Fermi had established his reputation with important work in theoretical physics. His most notable achievement in this area was his concept of radioactive beta decay. This concerned the theory that a proton could be created from a neutron via the shedding of an electron (a beta particle) and something known as an antineutrino.

It was for his achievements in experimental physics for which the Italian would be remembered, however, leaving behind a world, following his early death from cancer, very different from the one he had entered just over half a century earlier.

NUCLEAR CHAIN REACTION

γ

- U-235
- Ba-141
- Kr-92
- Proton
- Neutron
- γ Gamma Ray

When a neutron strikes a uranium-235 atom, it begins a nuclear chain reaction.

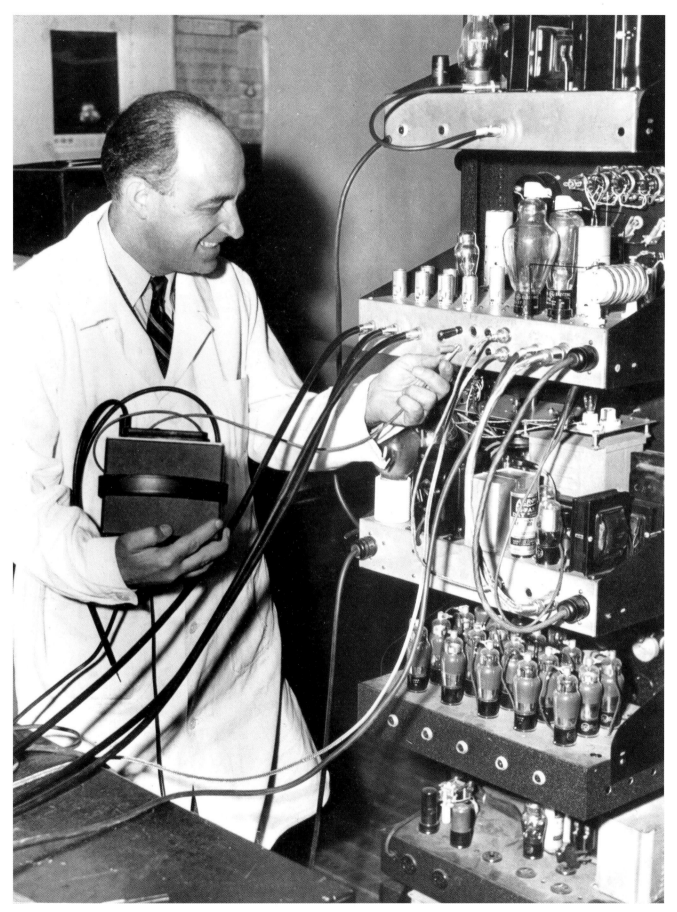

Enrico Fermi checks the circuit of a neutron counter in 1948.

WERNER HEISENBERG

GERMANY
1901–1976

IDEAS AND INVENTIONS
Matrix Mechanics, Uncertainty Principle, Copehagen Interpretation of Quantum Physics, Quantum Field Theory, Wave Function Collapse

FIELDS
Physics

Opposite: Werner Heisenberg.

'THE MORE PRECISE THE MEASUREMENT OF POSITION, THE MORE IMPRECISE THE MEASUREMENT OF MOMENTUM, AND VICE VERSA.'

- Werner Heisenberg

Heisenberg's development of matrix mechanics in 1925 sparked a controversy in the rarified world of quantum theory. Like many other physicists, Heisenberg too had been contemplating the debate over whether electrons and other atomic phenomena behaved in a wave or particle-like fashion. Heisenberg found a simple solution. He ignored both arguments altogether! Instead Heisenberg proposed that the only important factor was being able to mathematically predict the occurrence of atomic features which could be measured or observed such as frequency and light emissions. So, he applied algebra to the problem and developed a mathematically based solution which came to be known as matrix mechanics. The predictive and quantifiable powers of this new scheme were excellent and Heisenberg received the Nobel Prize for this development in 1932.

They also had a logical extension and it was to this part Einstein objected most. Heisenberg expressed it as his 'uncertainty' principle in 1927. In seeking to underline his basis for ignoring the visual idea of the atom and only considering it mathematically, Heisenberg realized in physical reality it was not possible to measure both the exact position and exact momentum of a particle at the same time. The reasoning behind this was simple: if one undertook an experiment to determine the position of, say, an electron at any instant, something like gamma rays would have to be deflected off the particle to locate it. In doing this, the position might be identified, but the electron's momentum would be radically changed by its interaction with the gamma rays. By the same token if less intrusive techniques were used to find

Werner Heisenberg writes an equation on the blackboard.

the electron, the original momentum might be better preserved, but the accuracy of the positioning would be woolly. This meant the best that could be hoped for was a mathematical prediction of the probability of the electron's position and location at any given instant, and Heisenberg supplied this formula.

Unfortunately, the logical conclusion from accepting this is that cause and effect as relied upon in classical physics can no longer be produced. The best that can be hoped for is a series of probabilities about the behaviour of any given particle at any point now or in the future. Max Born's 'probabilistic' interpretation, expressed at about the same time and concerning the likelihood of finding a particle at any particular point through probability defined by the amplitude of its associated wave, led to similar conclusions. On

Max Born was one of the leading advocates of the 'probabilistic' interpretation.

hearing of the radical ideas, Einstein remarked, 'God does not play dice. He may be subtle, but he is not malicious.' Still, the approach is now largely accepted.

Heisenberg partook in other important work, too. After Chadwick (1891–1974) had discovered the neutron in 1932, it was Heisenberg who suggested the model of the proton and neutron being held together in the nucleus of the atom. Moreover, Heisenberg played a significant and controversial role in Germany's attempts to develop an atomic bomb during World War II. Unlike many of his countrymen, Heisenberg did not leave his homeland when Hitler came to power, but by the same token he was no Nazi sympathizer. The government was well aware of Heisenberg's leading atomic knowledge, however, and compelled him to head up a team which would seek to create an atomic bomb. With the main Nazi focus being largely on developing other types of weapon, however, the team did not deliver in time to alter the course of the war. Furthermore, Heisenberg maintained after the war he had no intention of ever letting the project succeed and be in a position to hand over such a powerful device to Hitler anyway. He insisted, had it been necessary, he would have used his position to hijack the team's progress if it had come close to creating such a device.

Of all the competing models of quantum theory created in the 1920s, the theories developed by Werner Heisenberg, along with proposals based on similar principles by his German countryman Max Born (1882–1970), have endured the longest. The approach which turned physical science into merely a series of unpredictable probabilities appalled one of the greatest scientists of the century, Albert Einstein, amongst others, but Heisenberg's ideas worked and as a result they continued to be accepted.

A reproduction of the German Haigerloch nuclear project, which Heisenberg helped create.

LINUS
PAULING

USA
1901–1994

IDEAS AND INVENTIONS
Molecular Disease, Chemical Bonds, Protein Structures

FIELDS
Chemistry, Biology

Opposite: Linus Pauling.

'I WANTED TO UNDERSTAND THE WORLD.'

- Linus Pauling

Linus Pauling in his laboratory.

Linus Carl Pauling is particularly noted for his contributions to structural chemistry and his application of quantum theory in this area, as well as effectively founding molecular biology. In later life he achieved what only very few scientists ever succeed in, wide fame amongst the general public, principally for his anti-nuclear stance and advocacy of the health-giving properties of large quantities of vitamin C.

Pauling received his first Nobel Prize, for chemistry, in 1954, for the significant progress he had made in understanding chemical, in particular molecular, bonds. Earlier in the century, Pauling's American countryman Gilbert Lewis (1875–1946) had offered many of the basic explanations for the structural

bonding between elements which are now familiar in chemistry. These included the sharing of a pair of electrons between atoms and the tendency of elements to combine with others in order to 'fill' their electron 'shells', according to rigidly defined orbits (with two electrons in the closest orbit to the nucleus, eight in the second orbit, eight in the third and so on). Pauling now built on these efforts with research into more complicated bonds between molecules. Although he spent much of his career at the California Institute of Technology, the two years in Europe working alongside some of the finest brains in physical quantum theory in 1926 and 1927 greatly influenced his later work in structural chemistry.

He was one of the first to realize the impact of this new physical interpretation to understanding the bonds and nature of molecules and crystals from a chemical perspective. It was his application of quantum theory to structural chemistry, indeed, some say effectively founding structural chemistry in the modern sense, which enabled Pauling to make great strides. He went on to gather vast amounts of quantifiable data concerning the measurements and properties of molecules and crystals. This helped establish the subject and could be applied in making further predictions, including the formulation in 1929 of a series of influential rules about the stability of molecular structures. He summarized all his ideas in this area in his 1939 book *The Nature of the Chemical Bond and the Structure of Molecules and Crystals*, which became a leading authority.

Linus Pauling's triplex model of DNA, which was ultimately proved by the discoveries of Watson and Crick.

Pauling later moved into biochemistry again, effectively founding a new branch known as molecular biology through his discovery of the first 'molecular disease', sickle-cell anaemia. He also formulated theories on the immune system, provided chemical explanations for the way in which anaesthetics worked, and offered insights into the structure of proteins. He was also involved in the 'DNA race' to understand the nucleic acid's structure. Although his answer was ultimately wrong, Pauling provided context against which Watson (b.1928), Crick (1916–2004) and Wilkins (1916–2004) could compare their studies as well as take advantage of some of his methodologies.

Ironically, Pauling entered the public consciousness less for his chemical achievements and more for his stance against nuclear arms, and war in general. He refused to take part in the Manhattan Project during World War II and, indeed, his increasingly anti-nuclear stance after the war led to accusations of him being unpatriotic and to receiving some harassment from the authorities. For his efforts, though, he was awarded the Nobel Prize for Peace in 1962. Controversy continued to follow Pauling when he encouraged members of the public to take huge quantities of Vitamin C for its alleged health-giving properties. There was limited scientific evidence for this, but the turnabout it had brought in his own health was remarkable and provided the source of his advocacy.

Regarded by many as the most influential chemist of the twentieth century, Pauling is different from many of the scientists in this book in that he is not remembered for any single, specific world-changing theory, but more for a diverse range of improvements in biochemical understanding. With an incredible

Linus Pauling demonstrates in a protest against nuclear weapons.

memory, an instinctive feel for possible solutions to problems then confirmed by experimentation, and a talent for filling in the gaps between branches of science left behind by those working in specialized areas, he was the first person to receive two separate, individual Nobel Prizes.

Pauling's oxygen analyzer, which found use in the food, health and even space industries.

MIN CHUEH CHANG

CHINA
1908–1991

IDEAS AND INVENTIONS
Sperm capacitation, IVF, Oral Contraceptive Pill

FIELDS
Biology, Medicine

Opposite: Min Chueh Chang.

'SCIENTIFIC INQUIRY IS TO INCREASE THE WEALTH OF HUMAN KNOWLEDGE. THAT KNOWLEDGE CAN BE USED FOR GOOD OR BAD.'

- Min Chueh Chang

The work of reproductive biologist Min Chueh Chang contributed to unprecedented societal change. He jointly developed the first oral contraceptive pill, freeing millions of women from unwanted pregnancies. His insights into animal fertilization were milestones on the road towards human IVF (in vitro fertilization), which has enabled many with low fertility or reproductive disorders to have children themselves or via surrogacy.

Born in Taiyuan, northern China, Chang studied animal physiology in Beijing, then earned a PhD in animal breeding at Cambridge University. In 1951, he earned a fellowship at the Worcester Foundation for Experimental Biology (now part of the University of Massachusetts Medical School), where he learned the method for in vitro fertilization from its co-founder, Gregory Pincus.

In 1953, Chang and Pincus reported that the female reproductive hormone progesterone could prevent rabbits from ovulating (releasing eggs from the ovaries into the fallopian tubes for potential fertilization by sperm). This discovery had implications for the potential to control human fertility, leading the team at Worcester to develop a synthetic form of progesterone based on steroid compounds found in a type of yam. In 1960, it was licensed by the US government as the first combined oral contraceptive pill under the name Enovid.

Co-developing 'the pill' was Chang's highest-profile achievement, but his researches into animal fertilization, especially how eggs and sperm interact, was vitally important too.

Scientists first observed sperm fertilizing a female egg inside a living organism in the mid-19th century, and for the next 100 years attempts were made to replicate this phenomenon outside the body, or in vitro (literally 'in glass'). Gregory Pincus claimed to have engineered a successful birth by in vitro fertilization in 1934, but this was not verified.

It was Chang who made a major breakthrough in 1951, when he identified that, in rabbits, sperm needed to stay in the female's fallopian tubes for some time and undergo a certain biological process in order to fertilize the egg. This process was named capacitation by the Australian scientist Colin Austin, who made the same finding that same year. It was a turning point in the history of reproductive biology, allowing scientists to try to recreate the conditions for fertilization in the laboratory.

Over the next decade, research teams competed globally to make advances. Chang made another key discovery when he showed that sperm capacitation could occur in both the uterus and the fallopian tubes, and that it is a general phenomenon among mammals. Then, in 1958, reproductive biologists Anne McLaren and John Biggers at London's Royal Veterinary College reported the first birth of mice resulting from embryos grown in vitro. A year later, Chang performed what is widely recognized as the first unequivocal demonstration of a mammalian birth by IVF, taking eggs from a black female rabbit, fertilizing them in vitro using sperm from a black male, and placing them in a white female – which then gave birth to all-black young.

Chang's team continued researching the conditions required to achieve IVF in different mammals, reporting the IVF-assisted births of mice and rats and, in 1963, another first: the successful fertilization of golden hamster eggs using sperm that had been

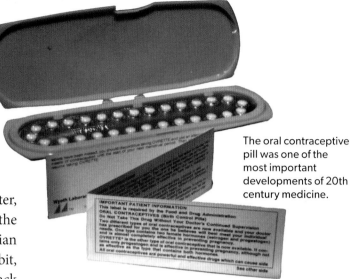

The oral contraceptive pill was one of the most important developments of 20th century medicine.

capacitated fully outside the female's body. These and other advances in animal fertilization by Chang and his associates are recognized as laying the foundations for human IVF and, from there, further assisted reproductive technologies such as vitrification (flash freezing) to preserve unfertilized eggs and the creation of ovarian tissue for future use.

HUMAN IVF AND BEYOND

In August 1978 Louise Brown was the first human born as a result of IVF. Further advances enabled researchers to grow embryos until the critical blastocyst stage of the reproductive process, then implant only the most robust embryo into the host. The resulting supply of 'excess' embryos gave rise to the field of human embryo research. Because of this, potential parents can now pre-screen embryos in advance for diseases and medical conditions, and scientists are able to switch off 'faulty' genes. The removal of stem cells from excess embryos for research is a long-established process, with doctors attempting to use those cells to repair damaged organs and tissue, or even grow new cells and organs for donation via cloning. The possibility of using human embryo research to eliminate certain illnesses or conditions, or to create 'designer' babies, raises a host of ethical and legal issues that practitioners in the field continue to wrestle with.

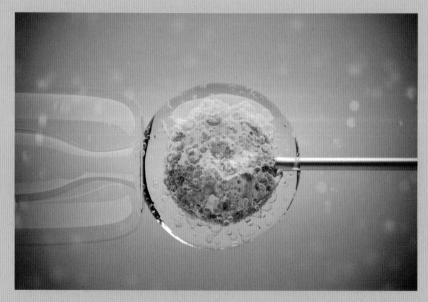

A 3D rendering of the process of IVF in humans.

Chang worked closely with Gregory Pincus in his research on reproduction.

WERNHER VON BRAUN

GERMANY
1912–1977

IDEAS AND INVENTIONS
V2 Rocket, Saturn V Rocket

FIELDS
Engineering

Opposite: Wernher von Braun.

'IT TAKES SIXTY-FIVE THOUSAND ERRORS BEFORE YOU ARE QUALIFIED TO MAKE A ROCKET.'

- Wernher von Braun

One of the most controversial and influential scientists of the 20th century, Wernher von Braun began his career in Nazi Germany by designing missiles to bomb London during World War II and ended it by creating the rocket that sent the first men to the Moon.

Born in 1912 in the eastern German city of Wirsitz (now Wyrzysk in Poland) the space-obsessed von Braun set up his own rocket design company in 1932, when he was barely out of his teens. Within a year he was conducting research for the German Army. With the Nazis in power from 1933, von Braun's career was always overshadowed by his association with fascism. He later claimed that he was compelled to join both the Nazi party and the SS in order to carry out his work, but just how much he objected to Adolf Hitler's National Socialism remains an open question.

Among those who inspired von Braun, foremost was Robert H. Goddard, the American who designed and launched the world's first liquid-fuelled rocket in 1926. Nicknamed 'Nell', Goddard's 3-m tall (10 ft) projectile was fuelled by gasoline and liquid oxygen and propelled itself 12.5 m (41 ft) into the air in a flight that lasted 2.5 seconds. Until Goddard's achievement, rocket-type objects (mainly fireworks and military ordnance) had been

Robert Goddard launches a liquid-fuelled rocket in 1926.

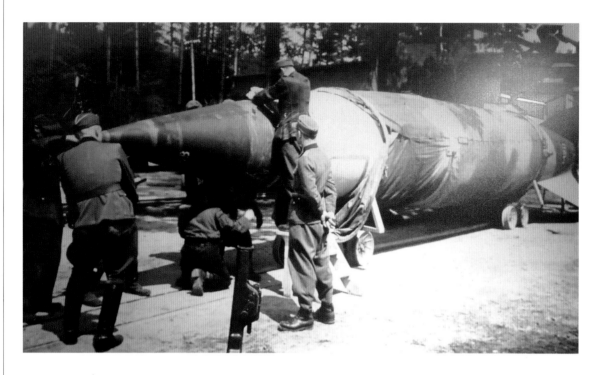

Wernher von Braun developed the V2 rocket, the first man-made object to reach space.

powered by gunpowder. Following Goddard, von Braun resolved to design bigger and better liquid fuelled rockets that could go into space and, one day, Mars. First, though, the German Army commissioned him to build rockets whose targets were closer to home.

As a technical director at the Peenemünde Army Research Centre on Germany's Baltic Coast during World War II, von Braun oversaw development of the notorious V-2 Vergeltungswaffe, or 'vengeance', rocket missiles launched at targets in Britain and Allied-controlled Europe from late 1944. Standing 14 m (46 ft) tall and carrying a 1 tonne (2,200 lb) warhead of TNT and ammonium nitrate, von Braun's V-2 rockets were powered by a mix of ethanol, water, and liquid oxygen. On 20 June 1944 von Braun achieved the first of his rocketry ambitions when a test V-2 crossed the 100 km-high (62 mile) Kármán Line, the boundary above the planet where Earth's atmosphere ends and the cosmos begins. For the first time ever, a man-made object had been sent into space.

At the end of World War II von Braun was one of 1,600 scientists taken to America from Nazi-occupied Europe as part of Operation Paperclip. This was a scheme that co-opted former enemy experts, some of whom, like von Braun, could have been tried for war crimes, into the new struggle against the Soviet Union in the Cold War. By 1950, von Braun was safely

in office at the US Army's Redstone Arsenal facility in Huntsville, Alabama. For the next 20 years this was the base from which he presided over America's ballistic missile and space rocketry programmes.

Early successes included key contributions to launching the first Western satellite, Explorer 1, in 1958, and making Alan Shepherd become the first American in space, on board the Mercury-Redstone spacecraft in 1961. But von Braun's greatest success was his input to the Apollo 11 mission that landed the

Von Braun stands next to the F1 engines of the Saturn V rocket.

first men on the Moon. It was von Braun's massive Saturn V rocket that got them there, its five cone-shaped F1 engines providing the 3.4 million kg (7.5 million lb) of thrust necessary to blast the 110-m tall (363 ft) spacecraft free of Earth's gravitational pull. It was fuelled by 1.2 million litres (318,000 gallons) of liquid oxygen and 770,000 litres (203,400 gallons) of kerosene.

Between its first test in 1967 and its last mission in 1973, 13 of von Braun's Saturn V rockets went into space, as part of the Apollo programme and to help build the Skylab space station. Von Braun was appointed NASA's Deputy Associate Administrator for Planning in 1970, but left two years later, frustrated that the organization's hierarchy and US politicians did not share his enthusiasm for more expansive space missions – particularly to his dream location, Mars. Diagnosed with cancer in 1973, Wernher von Braun died four years later, aged 65.

Right: The launch of Apollo 11, the mission that put the first men on the moon.

LIN
LANYING

'CAN SOMEONE GIVE ME ANOTHER TEN YEARS? IN TEN YEARS I CAN DEFINITELY FINISH WHAT I AM DOING AND I CAN DIE WITH NO REGRETS.'

- Lin Lanying

CHINA
1918–2003

IDEAS AND INVENTIONS
Semiconductors,
Microelectronics,
Monocrystalline Furnace

FIELDS
Materials Science, Engineering,
Physics

Above: Lin Lanying.

Most people think of semiconductors as tiny electronic devices, like microchips. More accurately, these devices are semiconductor units, as a semiconductor is the base material from which that device is made. Usually, that material is silicon, but other elements and compounds like germanium, gallium arsenide, and cadmium selenide are also semiconductors.

What makes a semiconductor special is its ability to both conduct an electrical charge and to regulate that charge so that its energy flows at an optimum level. For the first half of the 20th century, the Holy Grail of electrical engineering was to discover the ideal semiconductive material and develop a means by which it could be used to host advanced electrical components. One of the key people that achieved this was China's Lin Lanying. Her work transformed the electronics industry – not just on Earth, but in space.

Born in 1918, in the southern Chinese city of Putian, Lin was a clever child who fought hard to overcome her traditionally-minded family's objections to women's education. After earning a

BA in Physics from Fukien Christian University she remained there for another eight years, conducting research and teaching.

Lin left China in 1949 to study in America. A degree in Mathematics was followed in 1955 by a doctorate in Solid State Physics from the University of Pennsylvania – a feat no other Chinese national had achieved in a century. Lin took a job as a senior engineer with Sylvania Electric Products, arriving just in time to help its engineering team successfully manufacture the semiconductor monocrystalline silicon. This meant they had isolated single ('monocrystalline') crystals of the rock-like metalloid, producing a clean, pure, highly conductive surface onto which electrical components could be fixed.

This was an era defining breakthrough, one that helped to usher in the so-called Silicon Age, and the Computer and Digital Ages that followed. With the new-found ability to graft small but powerful and effective integrated circuits, or microchips, onto single crystals of silicon (the second most abundant element on the planet, after oxygen), electrical engineers were able to design ever smaller electronic devices. Within a few decades, for example, computers that once occupied entire buildings would be incorporated into hand-held mobile phone devices.

When she finally returned to China in 1957, Lin Lanying applied her knowledge and experience to China's own burgeoning electrical engineering and semiconductor industries. One of her areas of expertise was in synthesizing, essentially 'growing', crystals from germanium and silicon from which semiconductor units could be made. In 1962, she invented the monocrystalline furnace, a device able to mass-produce single crystals of pure, refined – and therefore more conductive – silicon for use in semiconductor units. This opened the door to microchip manufacture on an industrial scale. Today, 1.15 trillion semiconductor

Fukien Christian Campus University.

Monocrystalline silicon production in China. Lin Lanying developed a method for manufacturing the semiconductor – an essential breakthrough that helped spur the growth of the computing industry.

units are produced each year.

Having helped lay the foundation for semiconductor technology, Lin then demonstrated how a microchip's capacity could be further increased by a process known as epitaxy, where layers are added to individual silicon crystals. Silicon crystals are around the size of a thumbnail and are so thin that they can pass through the eye of a needle. In the earliest days of microchip technology, grafting a network of minuscule electrical components onto a crystal's surface was difficult work. Typically, a single crystal could only host a single transistor. Lin's technique of adding layers to a crystal meant that it could accommodate more transistors. Today, developments in micro- and nano-technology mean that one silicon crystal, or 'wafer', can contain around 30 layers and house billions of transistors.

From the mid-1970s Lin turned her attention to China's space programme, and in 1987 successfully devised a way to grow gallium arsenide semiconductor crystals in zero-gravity conditions, creating in effect extra-terrestrial microchips custom-made to work more efficiently in space. At a stroke, China's 'mother of semiconductor materials' also became its 'mother of aerospace materials'.

Throughout her long life, Lin Lanying was showered in official honours and sat on many prestigious national and international electrical engineering and semiconductor committees and boards. She never married and dedicated her life to her work. Lin Lanying was still actively pursuing research when she died of cancer on 4 March 2003, aged 85.

A motherboard made up of semiconductor units.

The launch of a Long March rocket from Jiuquan, China. Lin Lanying contributed to the Chinese space program from the 1970s onwards.

JAMES WATSON

USA
1928–

IDEAS AND INVENTIONS
Double Helix Model of DNA

FIELDS
Biology, Medicine

Opposite: James Watson.

'WE HAD FOUND THE SECRET OF LIFE.'

- James Watson

One of the most important scientific discoveries of the 20th century was made all the more interesting because the story behind it contained the key ingredients for the best drama: a race against time for a prize which would change the world, winners and losers, personality clashes, prejudice, a hint of sabotage, and a lingering question of 'what if?'

In fact, the story was so good, in 1968 James Dewey Watson, one of the main players, published it under the title *The Double Helix*. Rather than use the account to smooth over the passionate tensions and heat of the moment friction encountered at the time of its setting fifteen years earlier, he cranked up the drama another level with previously untold tales of ambition, conscious Nobel Prize pursuit, obstructive authority figures and spiteful side players. Perhaps the most unfair portrayal was that of Rosalind Franklin, relegated to a bit player when the crucial discovery was in fact hers (see box). It just added to the legend, of course.

The actual science behind the story, albeit in a slightly less dramatic way, is no less interesting. Early in 1953, the American Watson and his English colleague, Crick, announced that they had, quite literally, unravelled the secret of life. They had concluded that DNA, which was known to carry the hereditary information at the basis of all life, had a 'double helix' structure. Moreover, it was the detail of this construction which allowed it to pass on its secrets so successfully.

DNA INFOGRAPHICS

DNA BASE PAIRS

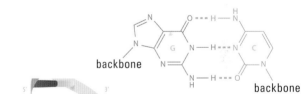

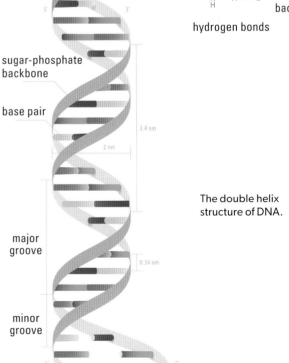

The double helix structure of DNA.

Rosalind Franklin

Rosalind Franklin pursued the task of figuring out the structure of DNA on her own. The link between DNA and its basis as the mechanism for passing on hereditary information had already been reasonably well established. The next step to understanding how DNA so successfully shared its data was to understand its structure. Franklin was an expert in X-ray diffraction techniques, a method which had been used for taking pictures of atoms in crystals and which was just starting to be applied to biological molecules. Franklin began examining DNA, then, via this means. The results of her investigations brought two important findings. Firstly, she realized that the 'backbone' of the molecule was on the outside, which Watson and Crick had at first missed, and was vital in eventually understanding its structure. Moreover, by 1952, Franklin had taken the clearest pictures of the molecules to date, which provided evidence of a helical, or spiral, structure. Watson and Crick would ultimately articulate a 'double helix' construction. However, Franklin's colleague Wilkins, who had access to her pictures, showed them to Watson. Immediately, Watson noted the evidence for the helical build, the vital piece in the jigsaw he and Crick had been trying to put together, and soon afterwards they made the breakthrough announcement that they had unravelled the structure of DNA. Watson's desultory description of her in his book on the discovery of the double helix helped to ensure that her role in the discovery was ignored. Shortly afterwards she died of cancer, aged just 37.

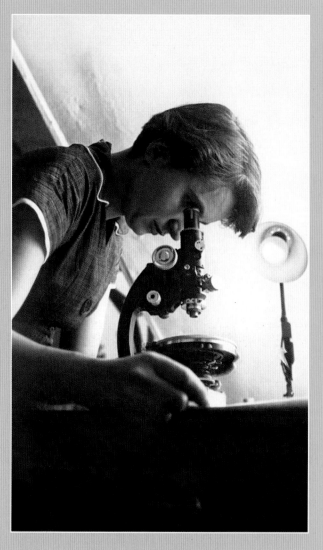

Key to DNA were the four bases adenine (A), cytosine (C), guanine (G) and thymine (T) which the scientist Erwin Chargaff (1905–2002) had earlier studied and measured. He had noted that C and G were always present in the same quantities, and A and T followed a similar pattern. Watson and Crick suspected this indicated some kind of mutual attraction between the respective bases, which meant they would only 'pair' with their appropriate partner within the backbone of a DNA molecule. They gradually fitted these ideas into a structure but still could not work out how the DNA molecule passed on its information so accurately. Then came the vital viewing by Watson of Franklin's X-ray diffraction photographs of DNA, secretly shown to him by Franklin's University of London colleague Maurice Wilkins (1916–2004). 'The instant I saw the picture my mouth fell open,' Watson later said. DNA was made in a helical structure, in fact as they would soon work out, a double helix, and over the following months Watson and Crick would finally realize why this was so important. When required to share its information, the two strands could literally uncoil themselves into two halves. This would leave the 'rungs of the ladder' containing A, C, G and T

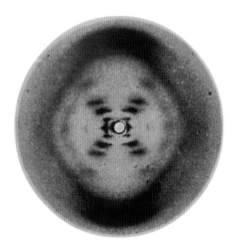

'Photo 51', the image taken by Franklin's graduate student of crystallized DNA, revealing a key clue to its structure.

exposed. Naturally, they would seek to 'complete' themselves again but as A would only link with T, C with G and vice versa, this meant the strands would combine selectively with other matter in the cell in order to form two perfect copies of its original self.

It was beautifully simple and the discovery made Watson and Crick world famous. Watson, Crick and, perhaps a little contentiously, Wilkins went on to collect the Nobel Prize for physiology for their discovery in 1962. By that point Franklin was already dead. But the controversy surrounding the tale, thanks especially to Watson's later book, was far from finished. The discovery of the double helix was the starting point from which scientific exploitation of DNA and genetic information would grow rapidly in the latter half of the twentieth century. Controversies and benefits surrounding genetically modified food, ethical dilemmas concerning cloning, and court cases hinging on DNA evidence would be some of the later developments dependent on this breakthrough.

The discovery of the double helix was the starting point from which scientific exploitation of DNA and genetic information would grow in the latter half of the twentieth century. Controversies surrounding and the possible benefits of genetically modified food, ethical dilemmas concerning cloning, and court cases hinging on DNA evidence would be some of the multitude of issues later spawned by this radical breakthrough.

In particular, Crick, Watson's colleague, would go on to make many additional contributions to the furthering of knowledge in the field.

Francis Crick.

WANGARI MAATHAI

KENYA
1940–2011

IDEAS AND INVENTIONS
Green Belt Movement

FIELDS
Environmental Science

Opposite: Wangari Maathai.

'UNTIL YOU DIG A HOLE, YOU PLANT A TREE, YOU WATER IT AND MAKE IT SURVIVE, YOU HAVEN'T DONE A THING.'

- Wangari Maathai

Some scientists change things because of the discoveries they make; others, like Kenya's Wangari Muta Maathai, make a difference because of their actions. As one of the earliest and most effective environmental activists in Africa, she overcame ethnic prejudice, government opposition, and misogyny to create a mass ecological movement, fight for democracy and women's rights, and win a Nobel Prize.

Born in 1940, Maathai's life changed when she went to America as part of the so-called 'Kennedy Airlift' of 1960. This was a scholarship scheme sponsored by then-senator John F. Kennedy that allowed 300 of Kenya's brightest young people to transfer to US universities. It was while studying in Pittsburgh for an MA in Biology that Maathai first encountered environmentalism, when activists there campaigned against air pollution. Initially, though, education and science were her priorities. In 1971, she was the first woman in East Africa to be awarded a PhD. She then became the country's first female senior lecturer in anatomy in 1975, the first female chair of a Kenyan university department (Veterinary Anatomy) in 1976, and the country's first female associate professor in 1977. But, as the decade progressed, environmentalism increasingly occupied her time, as did the fight for equal rights for women.

Both issues coalesced in the Green Belt Movement (GBM), the ecological organization Maathai established in 1977. The idea for the GBM came to Maathai when she joined the National Council of Women of Kenya (NCWK) and spoke with women in Kenya's countryside about their lives. She discovered that rural poverty and high incidences of disease and poor health were linked to environmental degradation: clean water sources were scarce and there was a severe lack of timber for construction, and for use in cooking and heating due to unregulated logging and forest clearance for animal grazing, mining, agriculture, and building. Maathai's brilliantly simple solution was to encourage Kenyans – particularly women – to plant more trees. To keep the process as natural as possible, GBM participants were encouraged to forage seeds from existing forests and plant them in treeless areas to establish new woodlands. The organization was supported with funds and expertise by the Norwegian Forestry Society and the United Nations Voluntary Fund for Women. This allowed the GBM to pay a small fee to its activists for every new tree they planted.

When Wangari Muta Maathai set up the Green Belt Movement, Kenya was a one-party state mired in corruption. GBM's work with the poor and marginalized showed Maathai and other progressively minded Kenyans how unjust their nation's politics were. In time, the GBM partnered with opposition

The forest in Aberdare National Park, Kenya. The Green Belt Movement project here led to the planting of nearly 4.1 million trees.

figures, pro-democracy campaigners, and pressure groups to demand change in Kenya. This led to Maathai's arrest, imprisonment, and physical assault on several occasions. Kenya's president, Daniel arap Moi, who took office in a rigged election in 1978, was an implacable opponent of Maathai. He publicly dubbed her 'a mad woman' and blocked her attempts at standing for public office. However, in the 2002 election that saw Arap Moi lose power – in no small part because of the pro-democratic advocacy of the GBM – Maathai was elected as an MP and a year later appointed Assistant Minister in the Ministry for Environment and Natural Resources.

When Wangari Muta Maathai was awarded the 2004 Nobel Peace Prize for her 'contribution to sustainable development, democracy and peace' she became the first African woman to receive the honour. It recognized a life devoted to environmental causes, women's rights, and democracy. She died in 2011 from complications of ovarian cancer, but her Green Belt Movement continues, in Kenya and in at least five other African nations. From its first public action on 5 June 1977, bedding in seven saplings in the Kenyan capital Nairobi's Kamakunji Park, hundreds of thousands of GBM activists have since planted 51 million trees in Kenya alone. This is in addition to the 1 billion trees planted globally in an initiative established in 2005 between Maathai and the United Nations Environment Programme (UNEP) – a scheme that now aims to plant another 13 billion trees.

Every tree planted as a result of Wangari Muta Maathai's pioneering work means less carbon dioxide in the atmosphere, a slower rate of global warming, more fertile soil that is less prone to desertification, and the growth of a valuable heating, cooking, and sustainable fuel resource for the people who need it most.

Wangari Maathai plants a tree in Uhuru National Park, Nairobi, Kenya, as (then) US Senator Barack Obama watches in 2006.

Daniel arap Moi, the dictator who ruled Kenya between 1978 and 2002.

FLOSSIE
WONG-STAAL

CHINA/USA
1946–2020

IDEAS AND INVENTIONS
HIV Cloning, Link Between HIV and AIDS

FIELDS
Biology, Medicine

Opposite: Flossie Wong-Staal.

'WORKING WITH THIS [HIV] VIRUS IS LIKE PUTTING YOUR HAND IN A TREASURE CHEST. EVERY TIME YOU PUT YOUR HAND IN, YOU PULL OUT A GEM.'

- Flossie Wong-Staal

The molecular biologist Flossie Wong-Staal was instrumental to the discovery that AIDS (acquired immunodeficiency syndrome) is caused by HIV (human immunodeficiency virus). She was the first scientist to clone the virus and identified how it worked at a molecular level. Her pioneering work in virology and immunology led to new diagnostic tools and successful therapies to manage HIV/AIDS, providing a model for investigating future viral threats such as COVID.

Born Yee Ching Wong in Guangzhou, southern China in 1946, Wong-Staal grew up in Hong Kong. Like many locals, she adopted an English first name. She studied at the University of California Los Angeles (UCLA), gaining a degree in bacteriology and a PhD in molecular biology. In 1973 Wong-Staal began a fellowship in the tumour cell biology laboratory of the National Cancer Institute, part of the US government's biomedical research agency, under biomedical scientist Robert Gallo. Her early research focused on areas including oncogenes (cells that have mutated and begin to rapidly divide and multiply) and the potential application to humans of animal retroviruses – a type of rapidly-multiplying virus able to change its genetic material into harmful DNA by invading host cells and integrating with its healthy DNA.

Although many scientists were sceptical about the existence of human retroviruses, in the late 1970s Robert Gallo identified the first one. Called HTLV-1, it was found in the T cells of leukaemia patients (a T cell being a type of infection-fighting white blood cell). With Gallo having identified the retrovirus, Wong-Staal did the gene sequencing and cloning work that showed it was carcinogenic. This marked

Robert Gallo identified the first human retrovirus.

her as a leading light in her field, and a key player among those who were at that time investigating a new disease accelerating through the gay community and infecting others, including blood donor recipients.

One thing that researchers identified fairly early on in the pandemic was the presence of abnormal T cells in AIDS patients. This led those studying it to suspect that AIDS was caused by a retrovirus. While it is widely accepted that the Pasteur Institute in Paris identified the HIV retrovirus first in 1983, it was Wong-Staal's expertise that proved pivotal in helping Robert Gallo's team identify a near-identical retrovirus soon after – and become the first researchers to convincingly show a causal link from HIV to AIDS. Overall, the work of the Pasteur Institute, Gallo's National Cancer Institute team, and another team of researchers at the University of California San Francisco (UCSF) enabled the first widespread blood screening programmes for HIV to be initiated in March 1985.

By now heading her own research team, Wong-Staal set out to uncover HIV's molecular makeup and its genetic methods of replicating. She achieved a landmark in 1985, cloning the virus for the first time. This provided additional confirmation of the link from HIV to AIDS: the clone virus that Wong-Staal created depleted host T cells in the same way that the actual virus had been shown to do in people with AIDS. Wong-Staal's further research into replication

then uncovered new HIV genes, including one known as VPR that helped to explain the virus's unusual ability to infect resting host cells as well as actively dividing cells. Wong-Staal's team also discovered genetic variations in HIV among patients and within the same patient. This revealed that the virus keeps mutating as the host immune system tries to fight it – meaning that creating a single, standard vaccine against it would not be effective. This is one of the reasons why, from the mid-1990s, 'cocktail' antiviral drug treatments were developed that were effective in managing the symptoms of AIDS and prolonging patients' lives. In addition, Wong-Staal's work also enabled the formulation of more sophisticated blood tests for detecting HIV.

In 1990 Wong-Staal was appointed as the Florence Seeley Riford Chair in AIDS Research at the University of California San Diego (UCSD). She went on to lead the university's prestigious Centre for Aids Research, carrying out investigations on the use of gene therapy to inhibit HIV replicating itself in stem cells, and offering some of the first AIDS-preventing gene therapy trials for HIV patients. In her later career Wong-Staal focused her expertise in HIV/AIDS towards researching improved treatments for hepatitis C, leading research at the biopharmaceutical company she co-founded until her death in 2002.

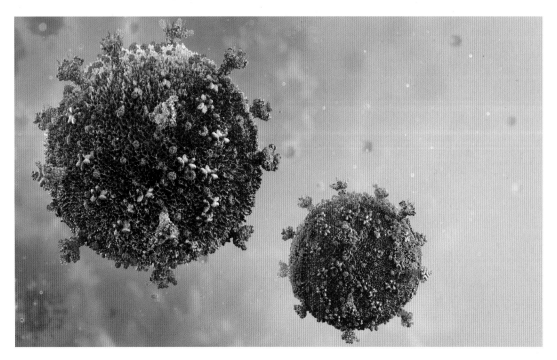

Left: The HIV virus. Flossie Wong-Staal identified the link between HIV and Aids.

Opposite: A researcher studies AIDS at the Pasteur Institute in Paris, France, in the 1980s.

CYNTHIA
KENYON

USA
1954–

IDEAS AND INVENTIONS
Genetic Role in Ageing

FIELDS
Biology

Opposite: Cynthia Kenyon.

'THE IDEA THAT AGEING WAS SUBJECT TO CONTROL WAS COMPLETELY UNEXPECTED.'

- Cynthia Kenyon

Few people would have imagined that the answers to the mysteries of human ageing would be found inside the genetic makeup of the humble worm. Fortunately for humankind, one of those few people was Cynthia Kenyon, the American molecular biologist whose pioneering studies in ageing research could make it possible for people to routinely live healthy and productive lives of 100 years or more.

Caenorhabditis elegans is an unremarkable nematode, or roundworm, that typically grows to around just 1 mm and has a life-cycle of 2–3 weeks. Kenyon first encountered it while conducting postgraduate research at Cambridge University in the UK, discovering that suppressing a particular growth gene in the worm and replacing it with the same gene from a fruit fly had no effect on the nematode's development. When she returned to America to take up a position as the University of California San Francisco (UCSF), Kenyon continued her investigations into the genetics of *C. elegans*. In 1993, and with the help of her student Ramon Tabtiang and others, Kenyon made her revolutionary breakthrough when she discovered that disabling a gene called daf-2 in the *C. elegans* nematode doubled its lifespan. Not only did the worm live longer; it suffered no ill effects or health issues either.

Before this discovery, ageing was viewed as an inevitable and progressive process of entropy, with every living thing ageing and declining at more or less the same rate. Kenyon showed that it was possible to intervene in the ageing process, to moderate it by deliberately mutating what Kenyon called 'the grim reaper' gene. By switching off the daf-2 gene in *C. elegans*, Kenyon and later researchers found that this provoked another gene, known as FOXO, to respond by turning on and off a number of other genes responsible for protecting and repairing tissues and organisms, actively helping them to survive for longer.

Studies show that humans with naturally-occurring mutations in their daf-2 and FOXO genes are more likely to live past 100. The challenge set by Cynthia Kenyon and other researchers into age is to create synthetically-manufactured mutations to these genes that work as well in humans as they do in the worms, flies and mice that they have successfully achieved this in so far. At the time of writing, they are not quite there, but if and when they do make it the result will not only mean longer lifespans for those whose genes are treated. During her experiments on *C. elegans*, Kenyon noticed that giving them small amounts of sugar shortened their lifespan by, as she told *The Guardian* newspaper in 2013, 'revving up the insulin pathway'. It's a complicated chemical process but Kenyon, and another scientist, Gary Ruvkun at Harvard Medical School, found that daf-2 controlled the worm's hormone-signalling system in the same

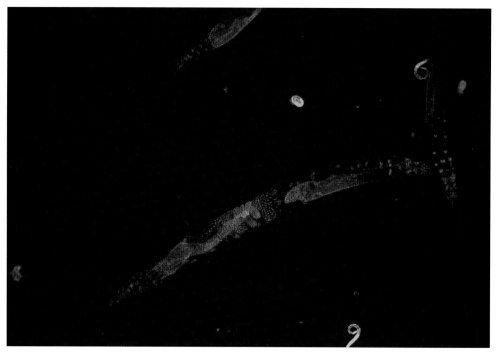

The *C. elegans* roundworm. By turning off one gene in this creature, Kenyon was able to significantly increase its lifespan.

way that the human body signals its need for the hormone insulin, which breaks down sugar, and the hormone IGF-1, which affects growth. In simple terms, this means that modifying the daf-2 gene could, one day, eliminate diabetes.

Much longer lives and an end to diabetes are not the only potential outcomes of Cynthia Kenyon's discoveries. By slowing down ageing and making it a healthier process, the result should be a massive reduction in age-related conditions such as cancer, heart disease, and Alzheimer's. Kenyon's later work on ageing also found that the daf-16 gene, along with daf-2, controlled an organism's antioxidant release. As antioxidants play a key role in moderating a body's immune system, this revelation indicates that modifying the daf-16 gene could have similar anti-ageing and life-

lengthening qualities as daf-2. Kenyon's work also inspired researchers in her field to discover that yet another gene, sir-2, has life-lengthening properties – and that adding an extra copy of that gene to nematodes doubles their life expectancy. More recent technical advances now mean that worms with modified genes can now live six times longer than normal.

Cynthia Kenyon is a multiple award winner. Her groundbreaking work could have truly far-reaching implications on human health and longevity if the successes in *C. elegans* gene modification can be replicated in us, *Homo sapiens*. Kenyon herself has declared her desire to live to at least 100 and, at the time of writing, is two-thirds of the way there and continues to actively work, speak and conduct research at UCSF and the privately funded Calico Labs in San Francisco.

Right: Kenyon's work may allow people to live much longer, healthier lives in the future and suffer less from an array of age-related maladies.

TIM
BERNERS-LEE

ENGLAND
1955–

IDEAS AND INVENTIONS
World Wide Web

FIELDS
Computer Science

Opposite: Tim Berners-Lee.

'THIS PROJECT ... COULD START A REVOLUTION IN INFORMATION ACCESS.'

- Tim Berners-Lee

Step back two and a half thousand years to the beginning of this book and consider for a moment the mysterious existence Anaximander (c. 611–547 BCE) was grappling with. Now fast-forward to the present day and attempt to comprehend how science has changed our understanding of the world and our successful manipulation of the elements contained within it. Tim Berners-Lee, the English inventor of the World Wide Web, is the last entry in a long line of those who have built on the work of others to change a facet of our lives, resulting in a modern existence which would be every bit as incomprehensible to Anaximander.

Little has changed our relationship with the world as much as the science of computers in which Berners-Lee works, and the World Wide Web has become a key tool in that revolution. Unlike equally significant recent developments, such as the joining together of a network of computers, a network of networks in fact, into an 'Internet', and the evolution of applications such as email which make use of this tool, the invention of the World Wide Web is particularly notable because it can be pinpointed to a sole creator, Berners-Lee. Rarely has the work of a single person had such a remarkable impact on business, research and individual lives as Berners-Lee's 1989 creation.

The World Wide Web is quite distinct from the Internet. The latter is the physical infrastructure through which data can be transmitted. The 'Web', however, was the first means by which the world at large gained access to and the ability to share information across this Internet.

The idea first occurred to Berners-Lee as an offshoot of a 1980 program he wrote called 'Enquire'. The concept was simple: he wanted to keep track of electronic information by 'linking' words in certain documents with other documents on his computer. Thus, Berners-Lee could jump from one related document, or piece of information, to the next, with minimum effort. Over the following years, the Englishman began considering ideas for allowing him to link to documents on other people's computers, and from theirs to his, without the need for a central database. It was as a logical extension of this vision that in 1989 he proposed a project to be called the World Wide Web.

Berners-Lee wrote a simple common language called Hypertext Mark-Up Language (HTML) through which authors could prepare documents in a common format with the necessary links, a method for linking these pages across the Internet (Hypertext Transfer Protocol or HTTP) and an addressing system for identifying and accessing the pages via a Universal Resource Locator (URL). His next step was then to create a straightforward Graphical User Interface (GUI) via which ordinary, non-technical people could read and share in these pages. It was launched onto the

Above: The first source code for the World Wide Web.

Internet at large in 1991, and in no time people were linking their pages across the world in a completely 'uncontrolled' Web.

Berners-Lee, like the rest of the world riding on the back of his creation, has come a long way since graduating from the University of Oxford in 1976. His early career in Dorset gave little insight into the revolution which would follow. He then went on to work in Geneva as a software consultant at the European laboratory for particle physics called CERN, where he had his initial idea for Enquire. Today, Berners-Lee lives and works in the United States at the Massachusetts Institute of Technology in its Computer Science Laboratory. He heads the World Wide Web Consortium, or 'W3C', which aims to 'lead the web to its full potential'.

Thus another application of science, the World Wide Web, brought about through the far-sighted vision of a single scientist, has again changed the world. How many more times will science continue to do so over the next two and a half thousand years? Our guess would probably be no better than Anaximander's.

Ongoing improvement in 'browsers', in particular, as well as other technology, have facilitated easier access to the Web to the point today where literally tens of millions of people make use of it every day, a number which is still growing. Now if we want to buy a car, research an essay, listen to the radio or find out a weather report, amongst thousands of other things, it can all be done on the Web in a way that was impossible even as recently as the mid-1990s.

Right: The first web server, used by Tim Berners-Lee at CERN in 1989. Two years later the web was opened up to the world at large.

JENNIFER
DOUDNA

USA
1964–

IDEAS AND INVENTIONS
RNA, CRISPR

FIELDS
Biology, Chemistry, Medicine

Opposite: Jennifer Doudna.

'THE POWER TO CONTROL OUR SPECIES' GENETIC FUTURE IS AWESOME AND TERRIFYING. DECIDING HOW TO HANDLE IT MAY BE THE BIGGEST CHALLENGE WE HAVE EVER FACED.'

- Jennifer Doudna

Every process in every organism's life is controlled by its genome – the unique genetic code comprising all of the DNA molecules in its cells, organized into sequences of information we call genes. In recent decades scientists have been developing methods to alter DNA, to remove genetic mutations or fight inherited diseases, for example. The outstanding figure in this area is American biochemist and structural biologist Jennifer Doudna who, in 2020, was given the Nobel Prize in Chemistry with the French microbiologist Emmanuelle Charpentier for, as their award citation said, 'the development of a method for genome editing'.

In her early career, Jennifer Doudna focused on RNA (ribonucleic acid) research. This was a molecule that was seen at the time as a mere message carrier for DNA (deoxyribonucleic acid), its role being to tell other cells when to begin making proteins. Doudna showed that RNA was actually able to initiate this process itself, and was a molecule worthy of further study. In 2005, as professor of Chemistry and Molecular and Cell Biology at the University of California Berkeley, this interest in RNA's potential led Doudna to the related field of CRISPRs (clustered regularly interspaced short palindromic repeats).

CRISPRs, in very simplified terms, are pieces of DNA that help an organism fight viruses. They do this by copying and storing parts of the DNA of a virus when it invades an organism, building an immune memory against it. If the virus attacks the organism again, the CRISPR DNA produces RNA molecules that recognize matching DNA in the virus and guide a defensive protein called Cas9 towards it. The Cas9 proteins are armed with 'scissors' that cut through the invading DNA and disable the virus.

Once it was known how CRISPRs worked, scientists began looking for ways to manipulate them. Chief among these was Jennifer Doudna who, along with Emmanuelle Charpentier, announced in 2012 that they had successfully adapted the CRISPR-Cas9 process to perform DNA microsurgery. They managed this by instructing the guide RNA molecules to look for specific DNA sequences in an organism, rather than ones only carrying viruses, and sent the Cas9's genetic scissors to cut them out. Doudna and Charpentier's work built on existing discoveries, and came at a

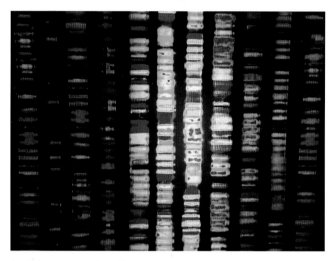

A DNA sequence. Jennifer Doudna played a key role in developing techniques in genetic sequencing.

time when others were making similar findings, but it was their research that won the most attention and appeared to have the best chance of practical application. What their gene editing breakthrough offered was the hope that CRISPRs could be used not just to fight viruses but for a host of other activities much wider and more ambitious in scope.

By harnessing the ability to modify the genome of plants, animals, and humans by editing their DNA, Doudna and Charpentier opened up new worlds of possibilities. Since their discovery, CRISPR has been used to prevent rice blast disease and to make tomato plants begin fruiting earlier. In animals, the technology has successfully eliminated muscular dystrophy, and has treated rare genetic mutations causing retinal dystrophy, liver failure, and lung fibrosis in mice. Longer term

research is looking into using CRISPR with stem cells to grow human organs in animals, and modifying insect DNA to eradicate malaria and Lyme disease.

The first US clinical trials on humans of CRISPR-based therapy began in 2019, to treat sickle cell disorder, and in 2020 CRISPRs were injected into six human patients with a condition called Leber congenital amaurosis that caused hereditary blindness. By the following year, three of the six had regained a small degree of vision. This followed developments in China where, in 2015, biologists reported the first use of CRISPR-Cas9 to modify human embryos. In 2018 another Chinese researcher announced the births of babies whose genes had been modified as embryos during in vitro fertilization (an application of CRISPR technology that Jennifer Doudna questions). On safer ethical ground, Doudna has conducted promising research on the potentially cancer-causing HPV (human papillomavirus infection) that indicates how her work could lead to giant steps forward in finding truly effective ways to prevent or cure cancer.

The potential of CRISPR technology is immense and it already appears to be a pivotal advance in science that could eliminate many illnesses and chronic medical conditions, and, in agriculture, provide hardier, more abundant, and longer lasting produce. Of all those working on CRISPR development, Jennifer Doudna at the time of writing remains perhaps the most important, influential, and active figure in this rapidly growing and ever-changing field.

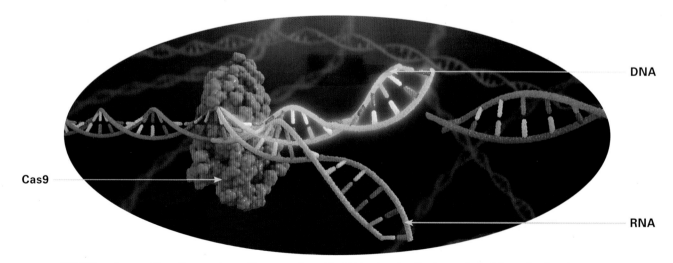

DNA

Cas9

RNA

CRISPR can be used to edit genes by making a precise cut in DNA using the Cas9 protein and then allowing the natural repair processes of DNA to mend the damage.

Emanuelle Charpentier worked alongside Jennifer Doudna in researching CRISPR.

PICTURE CREDITS

t = top, b = bottom, l = left, r = right